COLLOID FORMATION AND GROWTH
A Chemical Kinetics Approach

Colloid Formation and Growth
A CHEMICAL KINETICS APPROACH

JULIAN HEICKLEN

Department of Chemistry
The Pennsylvania State University
University Park, Pennsylvania

and

The Casali Institute of Applied Chemistry
The Hebrew University of Jerusalem
Jerusalem, Israel

ACADEMIC PRESS   New York San Francisco London   1976
A Subsidiary of Harcourt Brace Jovanovich, Publishers

COPYRIGHT © 1976, BY ACADEMIC PRESS, INC.
ALL RIGHTS RESERVED.
NO PART OF THIS PUBLICATION MAY BE REPRODUCED OR
TRANSMITTED IN ANY FORM OR BY ANY MEANS, ELECTRONIC
OR MECHANICAL, INCLUDING PHOTOCOPY, RECORDING, OR ANY
INFORMATION STORAGE AND RETRIEVAL SYSTEM, WITHOUT
PERMISSION IN WRITING FROM THE PUBLISHER.

ACADEMIC PRESS, INC.
111 Fifth Avenue, New York, New York 10003

United Kingdom Edition published by
ACADEMIC PRESS, INC. (LONDON) LTD.
24/28 Oval Road, London NW1

Library of Congress Cataloging in Publication Data

Heicklen, Julian.
    Colloid formation and growth.

    Bibliography: p.
    Includes index.
    1.   Colloids.   2.   Chemical reaction, Rate of.
I.   Title.
QD549.H46   541'.3451   75-19646
ISBN 0–12–336750–6

PRINTED IN THE UNITED STATES OF AMERICA

*This book is dedicated to those
who have made my life exciting:
my research associates and students
past, present, and future*

If I forget thee, O Jerusalem
Let my right hand forget its cunning
Let my tongue cleave to the roof of my mouth

CONTENTS

Preface ix
Acknowledgments xi
List of Symbols xiii

Chapter I. INTRODUCTION   1

Chapter II. FIRST-ORDER PHYSICAL LOSS PROCESSES   5

Diffusional Loss   5
Gravitational Settling   14
Evaluation of Parameters   15

Chapter III. BIMOLECULAR PROCESSES   21

Rate Coefficient   21
Condensation   26
Pure Coagulation   34

Chapter IV. THERMODYNAMICS AND REVERSE REACTIONS 40

Vaporization   40
Spontaneous Fracture   43
Supersaturation   45
Results at Low n 48

Chapter V. HOMOGENEOUS NUCLEATION   57

Equilibrium Theory   57
Steady-State Theory   63
Kinetic Theory   68
Effect of Diluent   71
Critical Supersaturation   72
Nucleation from Chemical Reaction   78

Chapter VI. HETEROGENEOUS NUCLEATION   87

Soluble Condensation Nuclei   87
Insoluble Condensation Nuclei   92
Ions as Condensation Nuclei   97

Chapter VII. ACCOMMODATION COEFFICIENTS   112

Transmission Factor $\Gamma_{n,m}$ 113
Steric Factor $p_{n,m}$ 115
Activation Energy $E_{n,m}$ 118
Comparison with Experimental Values   119
Summary   122

REFERENCES   123

INDEX   129

PREFACE

Over the years, certain facets of colloid science have raised a number of questions in my mind. These are:

Why don't billiard balls have sticky collisions?
Why don't water droplets divide like bacteria?
What happens to the surface tension and dielectric constant of a small drop as it evaporates and shrinks?
How important are wall reactions and gravitational settling?
When are rate coefficients diffusion controlled?
How big must a particle be before it is no longer in solution and is a separate phase?
How and how fast do particles nucleate from vapors?
Why does a small amount of an impurity ion promote crystallization from solution, but a large amount of the same ion retard crystallization?
What happens to a collection of coagulating particles?

Over the past few years we have initiated a research program on colloid nucleation and growth at Pennsylvania State University, and these questions have become even more nagging. Therefore this book was written to answer these and other questions.

In this book the science of colloid dynamics is developed from the viewpoint of chemical kinetics. I believe this represents an entirely new approach to this subject. The processes of homogeneous and heterogeneous nucleation, condensation, coagulation, vaporization, spontaneous fracture, diffusive loss, and gravitational settling are treated quantitatively. New derivations and results are included as well as a review of previous work. In some cases, simplified, easy-to-use, approximate formulas are developed for the various processes. Special attention is given to deviations for very small particles (i.e., clusters of ≤ 30 molecules), and a chapter is devoted to accommodation coefficients. Important features of the volume are:

Complete derivation of all rate coefficients for all processes of consequence in colloid dynamics.
Exact treatment of small particles (≤ 30 molecules).
Tables and figures of important results.
Discussion of theory of accommodation coefficients.
Review of previous work.

The scientific treatment is at the graduate student level. This book was developed for a one-quarter graduate course taught at the Casali Institute of Applied Chemistry, Hebrew University of Jerusalem. I hope it will be of value to all research workers and graduate students interested in environmental and colloid science.

ACKNOWLEDGMENTS

This book was drafted and many of the ideas developed while I was on sabbatical leave at the Casali Institute of Applied Chemistry at The Hebrew University of Jerusalem, 1973–1974. I extend my thanks to Professor Gabriel Stein, Director, and the staff of the Institute as well as the Physical Chemistry Department, for making my stay possible and fruitful. In spite of the fact that Israel was engaged in a traumatic war during my visit, and normal routines had to be readjusted, The Hebrew University and Israel continued to remain an exciting and interesting place, both professionally and personally.

In developing this book, I had useful conversations with Drs. R. G. de Pena, A. Glasner, and S. Sarig. Correspondence with Professors J. Kassner, Jr., and G. M. Pound and Dr. Welby Courtney was particularly beneficial. Of course, I drew heavily on other books and I mention the following as particularly helpful.

Amelin, A. G. (1967). "Theory of Fog Condensation" (English translation). S. Monson Wiener, Jerusalem.
Fuchs, N. A. (1964). "The Mechanics of Aerosols" (English translation). Pergamon, Oxford.
Hidy, G. M., and Brock, J. R. (1970). "The Dynamics of Aerocolloidal Systems." Pergamon, Oxford.
Hirth, J. P., and Pound, G. M. (1963) "Condensation and Evaporation, Nucleation and Growth Kinetics" *Prog. Mat. Sci.* **11**. Pergamon, Oxford.

LIST OF SYMBOLS

A	Basic molecular unit in foreign condensation nucleus
[A]	Concentration of A
$[A]_{vp}$	Equilibrium vapor pressure of A with A_∞
A_a	Foreign condensation nucleus consisting of molecules of A, where a is a positive integer
$[A_a]$	Concentration of A_a
A_0	Initial pressure of the chemical source reactant
A	Area of reaction vessel
A_n	Total surface area of the particles of type C_n
$A_{a\cdot n}$	Surface area of $A_a \cdot C_n$
$\Delta A_{n,m}$	Maximum increase in surface area as two particles C_n and C_m coalesce
$a_{n,j}$	Expansion coefficients in Eq. (51)
$B_{nj}\{t\}$	Expansion coefficients at time t in Eq. (60)
b_1, b_2, b_3	Coefficients in empirical diffusion Eq. (37)
C	Basic molecular unit in the particle
[C]	Concentration of C

$[C]_0$	Concentration of C at the distance of closest approach to a particle C_{n+1}
$[C]_{vp}$	Equilibrium vapor pressure of C with C_∞
C_n	Particle consisting of n molecules of C, where n is a positive integer
$[C_n]$	Concentration of C_n
$[C_n]_0$	Concentration of C_n at the distance of closest approach to another particle C_m (i.e., at $r_n + r_m$)
$[C_n]_\infty$	Bulk concentration of C_n
$[C_n\{t\}]$	Concentration of C_n at time t
$[C_n\{z\}]$	Concentration of C_n at height z
$[C_n\{R\}], [C_n\{R_0\}]$	Concentration of C_n at a distance R or R_0, respectively
C_q	Particle of critical size needed for growth; C_q is the particle whose vapor pressure corresponds to the prevailing pressure of C
C_s	Smallest size cluster of C molecules that is considered a particle
D_n	Diffusion coefficient for C_n
$D_n\{\text{gas kinetic}\}$	Diffusion coefficient as computed from gas kinetic theory for C_n
$D_n\{\text{Stokes}\}$	Diffusion coefficient as computed from Stokes' law for falling spheres for C_n
D_l'	$D_l P(2000/T)^{3/2}$
d	Distance between particles along their line of velocities
E	Expansion ratio
E_-	Expansion ratio needed for condensation on negative ions
E_+	Expansion ratio needed for condensation on positive ions
E_0	Expansion ratio needed for condensation when both positive and negative ions give the same results
E	Internal energy
E_n	Activation energy for $C_n \to$ physical loss
$E_{n,m}$	Activation energy for $C_n + C_m \to C_{n+m}$
e	Charge of the electron
F	Gibbs free energy
$\Delta F_{1,a\cdot n}$	Free energy change for the reaction $C + A_a \cdot C_n \to A_a \cdot C_{n+1}$
$\Delta F_{1,n}^0$	Free energy change at standard states for $C + C_n \to C_{n+1}$
$\Delta F_{1,\infty}^0$	Free energy change at standard states for condensation of C on an infinitely large particle C_∞

LIST OF SYMBOLS

$\Delta F_{1,n}^{\text{surf}}$	Free energy change at standard states due to surface energy for $C + C_n \to C_{n+1}$
$\Delta F_{1,a\cdot n}^{\text{dil}}$	Free energy change for dilution of C in $A_a \cdot C_n$
$\Delta F_{a,n}^{\text{surf}}$	Free energy change due to surface for the reaction $A_a + C_n \to A_a \cdot C_n$
$\Delta F_{a,n}^{\text{dil}}$	Free energy change of dilution for the reaction $A_a + C_n \to A_a \cdot C_n$
f_c, f_s	Correction function to R_v/R_w for infinite cylindrical and spherical reaction vessels, respectively; defined by Eqs. (13a) and (13b), respectively
g	Acceleration due to gravity = 980 cm/sec^2
H	Enthalpy
H_{vap}	Enthalpy of vaporization
ΔH_{vap}	Change in enthalpy of vaporization
ΔH_{soln}	Enthalpy of solution
ΔH_{adh}	Enthalpy of adhesion for a particle $A_a \cdot C_n$
h	Fall height
$I_0\{x\}, I_1\{x\}$	Modified Bessel functions of x of order 0 and 1, respectively
$J_0\{x\}, J_1\{x\}$	Zero-order and first-order Bessel function of x, respectively
$K_{n,m}$	Equilibrium constant for the reaction $C_n + C_m \rightleftarrows C_{n+m}$
K_{nucl}	$\prod_{j=1}^{q-1} K_{1,j}$
$K_{1,a\cdot n}$	Equilibrium constant for the reaction $C + A_a \cdot C_n \rightleftarrows A_a \cdot C_{n+1}$
K_{sp}	Solubility product
K_v	Defined as $k_n' R_0^2 / D_n$
K_w	Defined as $k_n\{\text{wall}\} R_0 / D_n$
k_n, k_{-n}	Rate coefficients for the forward and reverse reactions, respectively, of physical source-loss reactions: $C_n \rightleftarrows$ physical source
$k_n\{\text{eff wall}\}$	Effective net wall collision removal speed for species C_n; it is the velocity representing difference between wall deposition and wall disintegration rates
$k_n\{\text{wall}\}$	Net average speed at which C_n is removed at the wall
k_n'	Pseudo first-order rate coefficient for all the homogeneous loss terms for C_n (see Eq. (1))
$k_{n,m}, k_{-n,m}$	Rate coefficients for the forward and reverse reactions, respectively, of the coalescence–disintegration reactions $C_n + C_m \rightleftarrows C_{n+m}$
$k_{n,m}'$	$k_{n,m}(2000/T)^{1/2}$
$k_{1,a\cdot n}$	Rate coefficient for the reaction $C + A_a \cdot C_n \to A_a \cdot C_{n+1}$

$k_{1,a \cdot n+1}\{\text{eff}\}$	Effective net rate coefficient for condensation; it represents the difference of the condensation rate of C on $A_a \cdot C_{n+1}$ and the vaporization rate of $A_a \cdot C_{n+1}$ into $A_a \cdot C_n + C$
k_{coag}	Average coagulation rate coefficient
k_{cond}	Average condensation rate coefficient
k_{diff}	Rate coefficient for diffusional loss of active gas phase species (usually C)
k_{nucl}	Nucleation rate coefficient
k_0	Second-order rate coefficient for chemical source term $N\{0\}k_{\text{coag}}h\eta/\rho g r_{go}^2$
L	Length of a cylindrical reaction vessel
l	Length of a cube of a crystal
l_c	Cube root of the volume of one ion pair in a crystal
l_n	Mean free path of C_n in N_2
l_{N_2}	Mean free path of N_2 in N_2
M_n	Molecular weight of C_n
m_C	Mass of C
m_A	Mass of A
m_{N_2}	Mass of diluent gas molecule (i.e., N_2)
m_n	Mass of C_n
N or N_p	Total number density of particles; $N \equiv \sum [C_n]$
N_{max}	Maximum value of N
N_0 or $N\{0\}$	Initial value of N
$\mathcal{N}$	Avogadro's number $= 6.023 \times 10^{23}$
N_{N_2}	Number density of the diluent gas N_2
$n\{r\}$	Concentration of particles of radius r
P	Pressure
p_n	Steric factor for reaction $C_n \to$ physical loss
$p_{n,m}$	Steric factor for reaction $C_n + C_m \to C_{n+m}$
$p_{n,m}^\phi$	Part of $p_{n,m}$ due to angle of impact
$p_{n,m}^\sigma$	Part of $p_{n,m}$ due to orientation factors
Q	Rate of the chemical source reaction: chemical source $\to$ C
Q_{zz}	Elements of the diagonalized quadrupole tensor
q	Dipole moment
R_{nucl}	Rate of nucleation
$R_{n/m}$	Rate of reaction of particles C_n with one molecule of C_m
R_n'	Rate at which C_n is produced by coalescence of smaller particles and disintegration of larger particles. (For $n = 1$, R_1' also includes production from the chemical source terms Q)

LIST OF SYMBOLS

R_v	Total rate of volume removal processes
	$R_v \equiv k_n' \int_{R=0}^{R=R_0} [C_n\{R\}] \, dV$
R_w	Total rate of wall removal processes
	$R_w \equiv A k_n \{\text{wall}\} [C_n\{R_0\}]$
$R\{x\}$	Rate of production of x
R	Arbitrary distance from center of a spherical reaction vessel or from axis of cylindrical reaction vessel
R_0	Radius of reaction vessel
r_{N_2}	Radius of diluent N_2 molecules – computed from gas viscosity measurements
r_n	Radius of C_n
r_1	Effective equivalent spherical radius of C as determined from gas viscosity measurements
r_C	Effective equivalent spherical radius of C based on macroscopic density of C_∞
r_A	Effective equivalent spherical radius of A in a foreign condensation nucleus A_a
$r_{a\cdot n}$	Radius of particle of $A_a \cdot C_n$
$\bar{r}_1$	Average value of the molecular radius of $A_a \cdot C_n$
	$\bar{r}_1 \equiv (ar_A^3 + nr_C^3)^{1/3} / (n+a)^{1/3}$
$r'_{a\cdot n}$	Effective coulombic radius for $A_a \cdot C_n$
r_g	Geometric mean radius of a collection of particles
r_{g0}	Initial mean geometric radius of a collection of particles
r	Variable distance between any two particles C_n and C_m
S_m	Supersaturation of C_m, defined by $S_m \equiv [C_m]/[C]_{vp}$
$S_{m,m+n}$	Steady-state supersaturation of C_m with respect to C_{m+n}, i.e., supersaturation when deposition rate of C_m on C_{n+m} equals rate of disintegration of C_{m+n} into $C_m + C_n$
$S_{1,a\cdot n}$	Steady-state supersaturation of C with respect to $A_a \cdot C_n$
$S^0_{1,a\cdot n+1}$	Steady-state supersaturation of C with a drop composed only of C but with the same radius as $A_a \cdot C_{n+1}$
S	Entropy
T	Temperature
t	Time
V_n	Volume of a particle C_n
$V_{a\cdot n}$	Volume of a particle $A_a \cdot C_n$
V_a	Volume of particle A_a
V	Volume of reaction vessel

LIST OF SYMBOLS

V_z	Molar volume of gas at height z
V_0	Molar volume of gas at zero height
v_n	Speed of C_n
$\bar{v}_n$	Average speed of C_n
v_g	Settling velocity
X	Characteristic length associated with diffusion to the walls of the reaction vessel
x	Variable parameter
Z_n	Gas kinetic collision rate coefficient between C_n and diluent N_2
$Z_{n,m}$	Gas kinetic collision rate coefficient between C_n and C_m
$Z_{n,n+m}\{\text{eff}\}$	Effective net collision rate coefficient for coalescence; it represents the difference of the coalescence rate of C_n with C_{n+m} and disintegration rate of C_{n+m} into $C_n + C_m$
$Z_{1,a \cdot n}$	Collision rate coefficient between C and $A_a \cdot C_n$
z	Height
α_n	Accommodation coefficient for C_n reacting with the wall
$\alpha_{n,m}$	Accommodation coefficient for reaction between two particles C_n and C_m
α	Molecular polarizability
$\beta_{a \cdot n + 1}$	Defined by Eq. (199)
Γ_n	Transmission factor for $C_n \to$ physical loss
$\Gamma_{n,m}$	Transmission coefficient for the reaction: $C_n + C_m \to C_{n+m}$, i.e., the fraction of reaction which forms thermally equilibrated C_{n+m}
$\Gamma_{n,m}^p$	Part of $\Gamma_{n,m}$ due to physical restrictions
$\Gamma_{n,m}^T$	Part of $\Gamma_{n,m}$ due to thermal restrictions
$\Gamma\{x\}$	Gamma function of x
γ	Surface tension
γ_n	Surface tension for C_n
γ_∞ or γ_C	Macroscopic surface tension for C_∞
γ_A	Macroscopic surface tension for A_∞
$\gamma_{a \cdot n}$	Surface tension of $A_a \cdot C_n$
$\bar{\gamma}_{a \cdot n}^\infty$	Average surface tension for $A_a \cdot C_n$ $\bar{\gamma}_{a \cdot n}^\infty \equiv (a\gamma_A + n\gamma_C)/(a + n)$
δ	Tolman parameter for surface tension (see Eq. (104))
ε_∞	Macroscopic dielectric constant of C

LIST OF SYMBOLS

ε_0	Dielectric constant for the diluent gas N_2
$\varepsilon_{a \cdot n}$	Dielectric constant of $A_a \cdot C_n$
ζ_D	Correction factor for the diffusion coefficient as computed by simple gas kinetic theory to obtain exact result (see Eq. (33))
ζ_η	Correction factor for viscosity as computed by simple kinetic theory to obtain exact result (see Eq. (31))
η_{N_2}	Viscosity of diluent N_2
Θ_{nj}^2	$\mu_2\{n,j\}/\mu_0\{n,j\} - (\tau'_{nj})^2$
θ	$k_0/T^{1/2}$ (M^{-1} sec^{-1} °K$^{-1/2}$)
θ_n^2	$\sum_{j=2}^{n}(1/k'_{1,j})^2$
κ	Boltzmann constant = 1.38×10^{-16} erg/°K-molecule
Λ_j	$2k_{1,1}[C]^2/\omega_j J_1\{\omega_j\}$
λ	Potential energy interaction constant for $\Delta H_{adh} = \lambda/r_{a \cdot n}^\nu$
μ_n	Reduced mass of C_n and N_2; $\mu_n \equiv m^*_{N_2}m_n/(m_{N_2} + m_n)$
$\mu_{n,m}$	Reduced mass of C_n and C_m; $\mu_{n,m} \equiv m_n m_m/(m_n + m_m)$
$\mu_{1,a \cdot n}$	Reduced mass of C and $A_a \cdot C_n$
$\mu_i\{n,j\}$	$\int_0^\infty t^i a_{n,j}\{t\}\,dt$
$\mu_0'\{n,j\}$	$\prod_{l=1}^{n} k'_{1,l}/[k'_{1,l} + (D_l'\omega_j^2/[C]PR_0^2)T/2000]$
ν	Exponent of $r_{a \cdot n}$ in the interaction potential energy: $\Delta H_{adh} = \lambda/r_{a \cdot n}^\nu$
ξ	Time parameter $\equiv t[C](T/2000)^{1/2}$ M sec for constant $[C]$ or $t(A_0/8.1 \times 10^{-6}\,M)(T/2000\,°K)^{1/2}$ sec for variable $[C]$
ρ	Gas density
σ	Geometric standard deviation of the geometric mean radius of a collection of particles
σ_i or σ_0	Initial value of σ
τ_n	Induction time for the appearance of C_n
τ_n'	$\sum_{j=2}^{n} 1/k'_{1,j}$
τ'_{nj}	$\mu_1\{n,j\}/\mu_0\{n,j\}$
Υ	Geometric factor in Table II
Υ'	Correction factor in Eq. (204)
$\Phi_{n,j}$	$[C]^2(T/2000)\Theta_{nj}^2$
ϕ	Arbitrary angle
ϕ_{nj}	$[C](T/2000)^{1/2}\tau'_{nj}$
χ	$T/PA_0R_0^2$ °K/atm M cm^2
Ψ	Maximum fractional area increase during the reaction: $C_n + C_m \rightarrow C_{n+m}$

Ψ_{nj}^2	$[C]^2(T/2000)\Theta_{nj}^2$
ψ_m	Total flux of C_n through a spherical area toward one particle of C_m
Ω	$T/[C]PR_0^2$ °K M^{-1} atm^{-1} cm^{-1}
ω_j	Positive roots of the zero-order Bessel function $J_0\{\omega_j R_0\} = 0$
$\boldsymbol{\omega}_j$	Values of $\omega_j R_0$ for which $J_0\{\omega_j R_0\} = 0$

CHAPTER I

INTRODUCTION

The study of particle formation and growth is an old one, and many aspects of the problem are well understood. Most of the theory was developed by physicists and fluid dynamicists. Certain aspects of the problem are not yet clear, particularly as regards particle nucleation. In this treatise the science will be developed from the point of view of chemical kinetics, a viewpoint not previously considered in detail, as far as the author knows. Many derivations and the conclusions from them will, of course, be the same as already developed. In other cases the derivations may be different, but the conclusions the same. In still other cases different formulations will lead to different results.

We shall consider that there are the following fundamental processes that can occur:

$$\text{chemical source} \to C, \quad \text{rate} = Q$$

$$C_n \rightleftarrows \text{physical source}, \quad k_n; k_{-n}$$

$$C_n + C_m \rightleftarrows C_{n+m}, \quad k_{n,m}; k_{-n,m}$$

The first reaction involves some *in situ* chemical production of a molecule C which is the fundamental unit of the ultimate particle C_n which contains n such molecules. The rate of this production reaction we call Q, and we assume that the reaction is irreversible. In principle the reaction can be reversible, but for most systems of interest the reverse reaction is sufficiently unimportant that, for simplicity, it can be neglected. Examples of the species C and the corresponding chemical source term might be a H_2SO_4 molecule produced from the reaction of SO_3 and H_2O vapor, a polymer molecule produced from homogeneous polymerization of a soluble monomer, or a carbon atom produced from carbon suboxide photolysis.

The second reaction encompasses all physical means of removal of the particle C_n, where n is any positive integer. This includes flow out of the system (convection losses), diffusion losses (usually to the walls of the reaction system), and gravitational settling. The reverse reaction can also occur, and it corresponds to a physical production term, e.g., flow into the system, vaporization off the walls, or gravitational settling from above into the system. The rate coefficient for the forward reaction is designated k_n and that for the reverse reaction, k_{-n}.

The third reaction represents second-order coalescence terms between any two particles C_n and C_m (n or m can be any positive integer). The reverse reaction is disintegration. The forward rate coefficient is $k_{n,m}$ and the reverse rate coefficient is $k_{-n,m}$.

Certain types of reaction have been omitted in the preceding scheme. Of the two-body reactions, the following disproportionation reaction can occur:

$$C_n + C_m \rightleftarrows C_{n+l} + C_{m-l}$$

This reaction is almost equivalent to coalescence followed by disintegration

$$C_n + C_m \rightarrow C_{n+m}$$

$$C_{n+m} \rightarrow C_{n+l} + C_{m-l}$$

It differs only in that the intermediate C_{n+m} still contains the energy of coalescence, i.e., it has a higher effective temperature than the bath temperature and therefore a higher rate coefficient for disintegration. Except for impingment of C_n and C_m at high velocities, however, the disintegration of the energetic intermediate usually will not compete effectively with thermalization by the bath molecules. Once C_{n+m} is thermalized there is no distinction between the disproportionation reaction and the successive pair of coalescence and disintegration reactions just given. We will not consider the disproportionation reaction further in this book.

Reactions involving the simultaneous collision of three or more species are highly improbable and can never compete with reactions involving only two bodies. All third- and higher-order reactions can be treated as combinations of two-body reactions.

The problem that concerns us is to relate the fundamental reactions with observations and to deduce the rate coefficients. The reactions that are usually associated with particle formation and growth can be considered to be of the following five types:

$$\text{chemical source} \to C$$

$$C_n \rightleftarrows \text{physical source}$$

$$(q + 1)C \to C_{q+1}$$

$$C + C_n \rightleftarrows C_{n+1}, \quad n \geq q$$

$$C_n + C_m \rightleftarrows C_{n+m} \quad n,m \geq q$$

The first two reactions involving the chemical and physical source terms are the same as the fundamental reactions considered previously. The third reaction is the nucleation reaction. It involves a critical number $q + 1$ of vapor (or solution) phase molecules to unite to form a species C_{q+1} that will grow, where C_q is the critical size nucleus, i.e., one that is as likely to grow as to vaporize (dissolve). Particles smaller than C_q tend to revert to the molecular species C. Thus the system consists mainly of C and particles larger than C_q. As a result the coalescence–disintegration reactions are normally split into the last two of the preceding reactions. The reaction of C with any particle C_n ($n \geq q$) is condensation which leads to particle growth without change in particle number. The reverse reaction is vaporization. The coalescence reaction between C_n and C_m is referred to as coagulation when the two particles unite. The reverse disintegration process is spontaneous fracture.

In the foregoing model, the species C_n for $2 \leq n < q$ do not appear. Part of our task then is to justify their absence; another part is to determine the value of q, measure the rate coefficient for the nucleation reaction, and explain the apparent many-body interaction necessary for nucleation.

For convenience we here list all the reactions that we need to consider for a system involving one molecular species:

1. *Chemical source term* Chemical reaction which produces the fundamental molecule C

$$\text{chemical source} \to C$$

2. *Physical removal and source terms* Of most concern to us in this book will be *wall deposition and decomposition*, which are the forward and reverse reactions, respectively, for

$$C_n \rightleftarrows \text{wall}$$

Of these there are two special subclasses:
    a. wall condensation and vaporization, which are the forward and reverse reactions, respectively, for removal or formation of C at the wall

$$C \rightleftarrows \text{wall}$$

    b. wall coagulation and wall disintegration, which are the forward and reverse reactions, respectively, for removal or formation of particles $\geq C_q$ at the wall

$$C_n \rightleftarrows \text{wall}, \qquad n \geq q$$

3. *Nucleation* Conversion of several fundamental molecules into a stable product

$$(q + 1)C \rightarrow C_{q+1}$$

4. *Coalescence and disintegration* The joining of two particles into one and the splitting of one particle into two, respectively,

$$C_n + C_m \rightleftarrows C_{n+m}$$

Of these there are two special subclasses:
    a. Condensation and vaporization, which are the forward and reverse reactions, respectively, for removal of C by a species $\geq C_q$,

$$C + C_n \rightleftarrows C_{n+1}, \qquad n \geq q$$

    b. Coagulation and spontaneous fracture, which are the forward and reverse reactions, respectively, for removal of two particles each $\geq C_q$ into one larger particle and formation of two particles each $\geq C_q$ from one particle

$$C_n + C_m \rightleftarrows C_{n+m}, \qquad m,n \geq q$$

CHAPTER II

FIRST-ORDER PHYSICAL LOSS PROCESSES

In this chapter we discuss the two physical loss problems of kinetic interest: diffusional loss to the walls of the reaction system and gravitational settling.

Diffusional Loss

The general equation of concentration variation which must be considered is

$$\frac{d[C_n]}{dt} = D_n \nabla^2 [C_n] + R_n' - k_n'[C_n] \tag{1}$$

where t is the time, D_n the diffusion coefficient for C_n, and R_n' the rate at which C_n is produced at any point in the reaction vessel by coalescence of smaller particles and by disintegration of larger particles. If $n = 1$, i.e., for C, the chemical production term Q is also included in R_1'. The rate coefficient k_n' represents the pseudo first-order loss processes of C_n at any point in the interior volume of the reaction vessel, i.e., loss through reaction of C_n with another particle or by disintegration. In terms of the fundamental reactions

$$R_n' \equiv \sum_{j=1}^{n-1} k_{j,n-j}[C_j][C_{n-j}] + \sum_{j>n}^{\infty} k_{-n,j-n}[C_j], \qquad n > 1$$

$$R_1' \equiv Q + \sum_{j=2}^{\infty} k_{-1,j-1}[C_j], \qquad n = 1$$

$$k_n' \equiv \sum_{j=1}^{\infty} k_{n,j}[C_j] + k_{n,n}[C_n] + \sum_{j=1}^{n-1} k_{-j,n-j}, \qquad n \geq 1$$

Equation (1) must be solved with the following boundary conditions: $[C_n]$ is finite everywhere in the reaction vessel, and

$$\frac{-D_n d[C_n]}{dR} = k_n\{\text{wall}\}[C_n] \qquad \text{for} \quad R = R_0$$

where R is the distance from the center of the vessel and R_0 the radius of the vessel. The coefficient $k_n\{\text{wall}\}$ has the dimensions of velocity and represents the net average speed at which particles C_n are removed at the wall. It is a combination of three terms. These terms are

    1. the average speed at which the particles hit the wall,
    2. the accommodation coefficient α_n; i.e., the fraction of particles that hit the wall that are removed from the system (the fraction of particles that do not bounce), and
    3. the rate at which particles are emitted from the wall by vaporization of the wall.

In this chapter we assume that $\alpha_n = 1$ and that vaporization is negligible. Accommodation coefficients and wall vaporization will be considered later. Thus in this chapter we assume that the net rate of loss at the wall is equal to the rate at which particles hit the wall, i.e., the maximum rate possible.

The coefficient $k_n\{\text{wall}\}$ is then the average speed at which particles hit the wall. Consider particles of average speed $\bar{v}_n$ hitting a plane surface at

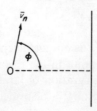

a zenith angle ϕ. The component of $\bar{v}_n$ that determines the speed of hitting the wall is that component of $\bar{v}_n$ perpendicular to the wall, i.e., $\bar{v}_n \cos \phi$. The fraction of particles having a velocity between the angles ϕ and $\phi + d\phi$ is that fraction of a sphere encompassed by the range ϕ to $\phi + d\phi$, i.e., $2\pi \sin \phi \, d\phi/4\pi$, where 4π represents the total solid angle of a sphere. Thus

$$k_n\{\text{wall}\} = \int_0^{\pi/2} (\bar{v}_n \cos \phi) \frac{(2\pi \sin \phi \, d\phi)}{4\pi} = \frac{\bar{v}_n}{4} \tag{2}$$

As far as the author knows Eq. (1) has not been solved in its most general form. Many special cases of the equation have been solved and solutions are available. In many of the solutions the boundary condition used was $[C_n] = 0$ at the walls of the reaction vessel. This condition is clearly incorrect, since it would require an infinite value for the coefficient $k_n\{\text{wall}\}$.

Simplified Solution

Since the general equation (1) has not been solved, special cases must be examined. The three terms on the right-hand side correspond to diffusional terms, homogeneous production processes, and homogeneous loss processes, respectively. Let us consider the contribution of the diffusional term:

$$\left(\frac{\partial [C_n]}{\partial t}\right)_{\text{diff}} = D_n \nabla^2 [C_n] \tag{3}$$

which represents physical loss processes through diffusion. (We are neglecting physical sources in this chapter.) In terms of the fundamental reactions discussed in Chapter I, we would like to represent the diffusional loss rate by

$$\left(\frac{\partial [C_n]}{\partial t}\right)_{\text{diff}} \simeq -k_n [C_n] \tag{4}$$

so that we can eliminate the spatial differential terms. Of course this transformation cannot be exact, but we can approximate it by making the transformation

$$-D_n \nabla^2 \Rightarrow k_n$$

This transformation changes the total wall loss rate into an average volume loss rate

$$k_n\{\text{wall}\}[C_n\{R_0\}]A = k_n \overline{[C_n\{R\}]} V, \tag{5}$$

where A and V are the area and volume of the reaction vessel, respectively, and $[C_n\{R_0\}]$ and $\overline{[C_n\{R\}]}$ are the wall concentration and average radial concentration of C_n, respectively. Equation (5) is the equation which defines k_n.

Now ∇^2 has the dimensions of reciprocal length squared. Thus we think of ∇ as representing the inverse of some characteristic length X, associated with diffusion to the walls, where X is of the order of the radius of the reaction vessel. Then

$$k_n = D_n/X^2 \tag{6}$$

In order to better estimate X, consider the ratio of homogeneous removal to wall removal processes

$$k_n'/k_n = k_n' X^2/D_n. \tag{7}$$

This ratio should be dependent on the ratio of volume to surface of the reaction vessel. Thus it seems reasonable to associate X with this ratio. If so, then, as an approximation:

For a cylindrical vessel of infinite length

$$k_n \simeq 4D_n/R_0^2 \tag{8a}$$

$$k_n'/k_n \simeq k_n' R_0^2/4D_n \tag{8b}$$

For a spherical vessel

$$k_n \simeq 9D_n/R_0^2 \tag{8c}$$

$$k_n'/k_n \simeq k_n' R_0^2/9D_n \tag{8d}$$

where $R_0/2$ and $R_0/3$ are, respectively, the ratio of volume to surface area for infinite cylindrical and spherical reaction vessels.

Steady-State Solution

The solution of Eq. (1) of importance in many chemical systems is when the steady state applies, i.e., when $d[C_n]/dt$ is negligibly small. Then Eq. (1) becomes

$$0 = D_n \nabla^2 [C_n] + R_n' - k_n'[C_n] \tag{9}$$

The solution to this equation was solved (Hudson and Heicklen, 1967) for the special case when R_n' and k_n' were constant throughout the reaction vessel, i.e., independent of R.

The problem was solved for both an infinitely long cylindrical reaction vessel and a spherical reaction vessel in terms of the dimensionless parameters

$$K_v \equiv k_n' R_0^2 / D_n$$

$$K_w \equiv k_n\{\text{wall}\} R_0 / D_n$$

Direct evaluation of Eq. (9) led to the results

For an infinite cylinder

$$\frac{D_n[C_n\{R\}]}{R_n' R_0^2} = \frac{1}{K_v} - \frac{(K_w/K_v) I_0\{K_v^{1/2} R/R_0\}}{K_v^{1/2} I_1\{K_v^{1/2}\} + K_w I_0\{K_v^{1/2}\}} \quad (10a)$$

For a sphere

$$\frac{D_n[C_n\{R\}]}{R_n' R_0^2} = \frac{1}{K_v} - \frac{(K_w/K_v) \sinh\{K_v^{1/2} R/R_0\}/(K_v^{1/2} R/R_0)}{\cosh\{K_v^{1/2}\} + (K_w - 1) \sinh\{K_v^{1/2}\}/K_v^{1/2}} \quad (10b)$$

where $I_0\{x\}$ and $I_1\{x\}$ are modified Bessel functions. The overall reaction rates are found from the expressions:

For an infinite cylinder

$$R_v = 2\pi L k_n' \int_0^{R_0} [C_n\{R\}] R \, dR \quad (11a)$$

$$R_w = 2\pi L k_n\{\text{wall}\} R_0 [C_n\{R_0\}] \quad (11b)$$

where L is the length of the cylinder;

For a sphere

$$R_v = 4\pi k_n' \int_0^{R_0} [C_n\{R\}] R^2 \, dR \quad (11c)$$

$$R_w = 4\pi R_0^2 k_n\{\text{wall}\} [C_n\{R_0\}] \quad (11d)$$

where R_v and R_w are, respectively, the removal rates via homogeneous and wall processes. The ratio R_v/R_w becomes:

For an infinite cylinder

$$R_v/R_w = (K_v/2K_w) + f_c \quad (12a)$$

For a sphere

$$R_v/R_w = (K_v/3K_w) + f_s \quad (12b)$$

where

$$f_c \equiv K_v^{1/2} I_0\{K_v^{1/2}\}/2 I_1\{K_v^{1/2}\} - 1 \quad (13a)$$

$$f_s \equiv (K_v/3)/(K_v^{1/2} \coth\{K_v^{1/2}\} - 1) - 1 \quad (13b)$$

Table I

Values of R_v/R_w for Steady-State Conditions

K_v/K_w	K_v						
	0.01	0.04	0.09	0.25	1.0	4.0	9.0	25

For an Infinite Cylinder

K_v/K_w	0.01	0.04	0.09	0.25	1.0	4.0	9.0	25
$<10^{-5}$ [a]	0.00125	0.0050	0.0113	0.0309	0.1201	0.4331	0.8519	1.798
2×10^{-5}	0.00126	0.0050	0.0113	0.0309	0.1201	0.4331	0.8519	1.798
2×10^{-4}	0.00135	0.0051	0.0114	0.0310	0.1202	0.4332	0.8520	1.798
2×10^{-3}	0.00225	0.0060	0.0123	0.0319	0.1211	0.4341	0.8529	1.799
2×10^{-2}	0.01125	0.0150	0.0213	0.0409	0.1301	0.4431	0.8619	1.808
2×10^{-1}	0.10125	0.1050	0.1113	0.1309	0.2201	0.5331	0.9519	1.898
2	1.00125	1.0050	1.0113	1.0309	1.1201	1.4331	1.8519	2.798
20	10.00125	10.0050	10.0113	10.0309	10.1201	10.4331	10.8519	11.798

For a Sphere

K_v/K_w	0.01	0.04	0.09	0.25	1.0	4.0	9.0	25
$<10^{-5}$ [b]	0.0007	0.00266	0.00598	0.0165	0.0648	0.2407	0.4889	1.083
3×10^{-5}	0.0007	0.00267	0.00599	0.0165	0.0648	0.2407	0.4889	1.083
3×10^{-4}	0.0008	0.00276	0.00608	0.0166	0.0649	0.2408	0.4890	1.083
3×10^{-3}	0.0017	0.00366	0.00698	0.0175	0.0658	0.2417	0.4899	1.084
3×10^{-2}	0.0107	0.01266	0.01598	0.0265	0.0748	0.2507	0.4989	1.093
3×10^{-1}	0.1007	0.10266	0.10598	0.1165	0.1648	0.3407	0.5889	1.183
3	1.0007	1.00266	1.00598	1.0165	1.0648	1.2407	1.4889	2.083
30	10.0007	10.00266	10.00598	10.0165	10.0648	10.2407	10.4889	11.083

[a] Entries in this row also give f_c.
[b] Entries in this row also give f_s.

DIFFUSIONAL LOSS

The functions f_c and f_s depend on K_v only, and they are tabulated in Table I for interesting values of K_v. They are identical to the values of R_v/R_w for $K_w/K_v < 10^{-5}$, and are the entries in the first row for the corresponding type of reaction vessel. Table I lists the values for R_v/R_w computed from Eqs. (12a) and (12b) for various values of K_v/K_w. R_v/R_w increases with K_v/K_w for a given value of K_v and increases with K_v for a given value of K_v/K_w. The table is terminated when the values of R_v/R_w are > 10, since then $> 90\%$ of the removal is by homogeneous processes, and diffusional loss is minor.

It is of interest to compare the exact values of $R_v/R_w = k_n'/k_n$ with the approximate values obtained from Eqs. (8b) and (8d). The ratios of the exact to approximate values of R_v/R_w are listed in Table II. For almost all chemical systems of interest in which diffusion plays any role, $K_w \geq 0.5$. For these conditions the approximate formula gives values within a factor of 4 of the exact formula. If $K_w < 0.5$, then serious deviations occur with the approximate formulas.

The limiting case of $R_v = 0$ also has special application, i.e., when there are no homogeneous removal steps. The rate constant for wall removal k_n is defined by Eq. (5), which rearranges to

$$k_n \equiv k_n\{\text{wall}\} A[C_n\{R_0\}]/V\overline{[C_n\{R\}]} \qquad (5')$$

where A and V are the area and volume of the reaction vessel, respectively, and $\overline{[C_n\{R\}]}$ is the average of $[C_n\{R\}]$ throughout the vessel. For very large vessels, $\overline{[C_n\{R\}]} \approx [C_n\{0\}]$; for other cases $[C_n\{0\}]$ will be an upper limiting value. If we replace $\overline{[C_n\{R\}]}$ by $[C_n\{0\}]$ in Eq. (5'), then a minimum value of k_n will be computed, but it will be approximately correct for most situations.

When $K_v = 0$, then

For an infinite cylinder

$$A/V = 2/R_0 \qquad (14a)$$

$$[C_n\{R_0\}]/[C_n\{0\}] = (1 + K_w)^{-1} \qquad (15a)$$

$$k_n \gtrsim 2k_n\{\text{wall}\}/R_0(1 + K_w) \qquad (16a)$$

For a sphere

$$A/V = 3/R_0 \qquad (14b)$$

$$[C_n\{R_0\}]/[C_n\{0\}] = 2/(2 + K_w) \qquad (15b)$$

$$k_n \gtrsim 6k_n\{\text{wall}\}/R_0(2 + K_w) \qquad (16b)$$

Table II

Values of $(R_v/R_w)(\Upsilon/K_v)$ for Steady-State Conditions[a]

K_v/K_w	K_v						
	0.01	0.04	0.09	0.25	1.0	4.0	9.0	25
	For an Infinite Cylinder							
$<10^{-5}$	0.500	0.50	0.502	0.495	0.480	0.433	0.378	0.288
2×10^{-5}	0.504	0.50	0.502	0.495	0.480	0.433	0.378	0.288
2×10^{-4}	0.540	0.51	0.506	0.496	0.481	0.433	0.378	0.288
2×10^{-3}	0.900	0.60	0.546	0.511	0.485	0.434	0.379	0.288
2×10^{-2}	4.50	1.50	0.945	0.655	0.520	0.443	0.383	0.289
2×10^{-1}	40.5	10.50	4.95	2.09	0.880	0.533	0.423	0.303
2	400	100.5	45.0	16.6	4.48	1.43	0.823	0.445
20	4000	1005	445	161	40.48	10.43	4.82	1.89
	For a Sphere							
$<10^{-5}$	0.63	0.60	0.60	0.594	0.583	0.541	0.489	0.390
3×10^{-5}	0.63	0.60	0.60	0.594	0.583	0.541	0.489	0.390
3×10^{-4}	0.72	0.62	0.61	0.598	0.584	0.541	0.489	0.390
3×10^{-3}	1.53	0.825	0.70	0.63	0.592	0.543	0.490	0.390
3×10^{-2}	9.63	2.85	1.60	0.954	0.672	0.564	0.499	0.394
3×10^{-1}	90.6	23.1	10.60	4.20	1.48	0.765	0.589	0.426
3	900.6	226	100.6	36.6	9.58	2.80	1.49	0.750
30	9006	2256	1006	360.6	90.58	23.0	10.49	3.98

[a] $\Upsilon = 4$ for an infinite cylinder and 9 for a sphere. Values under line correspond to $K_w \leq 0.5$.

Equations (16a) and (16b) have interesting limiting cases for small and large values of K_w:

For an infinite cylinder

$$k_n \gtrsim 2k_n\{\text{wall}\}/R_0, \qquad K_w \ll 1$$
$$k_n \gtrsim 2D_n/R_0^2, \qquad K_w \gg 1$$

For a sphere

$$k_n \gtrsim 3k_n\{\text{wall}\}/R_0, \qquad K_w \ll 2$$
$$k_n \gtrsim 6D_n/R_0^2, \qquad K_w \gg 2$$

The situation for K_w very small is the gas kinetic limit and k_n is independent of D_n. For K_w very large, the reaction is diffusion controlled, and k_n is independent of $k_n\{\text{wall}\}$.

Non-Steady-State Solution

The non-steady-state solution of the general equation (1) has been solved for the case of pure diffusional loss, i.e., when the homogeneous terms R_n' and $k_n'[C_n]$ are neglected, for many types of reaction vessels (Hidy and Brock, 1970, pp. 174–178). The equation to be solved is

$$\frac{d[C_n]}{dt} = D_n \nabla^2 [C_n] \tag{17}$$

For a spherical reaction vessel the boundary conditions considered were

$$[C_n\{R\}] = [C_n\{0\}] = \text{constant} \quad \text{at} \quad t = 0, \; R < R_0$$
$$[C_n\{R_0\}] = 0 \quad \text{at all times}$$

The second condition, of course, cannot be correct, but it simplifies the problem, and introduces significant error only near the wall of the reaction vessel. The solution of Eq. (17) is

$$[C_n\{R\}] = \frac{2R_0[C_n\{0\}]}{\pi R} \sum_{j=1}^{\infty} \frac{(-1)^j}{j} \sin\left\{\frac{j\pi R}{R_0}\right\} \exp\left\{\frac{-j^2\pi^2 D_n t}{R_0^2}\right\} \tag{18a}$$

For an infinitely long cylindrical reaction vessel and identical boundary conditions, the solution becomes

$$[C_n\{R\}] = \frac{2[C_n\{0\}]}{R_0} \sum_{j=1}^{\infty} \exp\{-D_n \omega_j^2 t\} \frac{J_0\{R\omega_j\}}{\omega_j J_1\{R_0 \omega_j\}} \tag{18b}$$

where $J_0\{x\}$ and $J_1\{x\}$ are, respectively, the zero-order and first-order Bessel functions, and the ω_j are the positive roots of $J_0\{\omega_j R_0\} = 0$. The solution for both reaction vessels underestimates the concentration of $C_n\{R\}$ because of the erroneous boundary condition.

Gravitational Settling

If, in addition to Brownian motion (resulting from molecular diffusion), the reaction system is under the influence of a force field, the particles will show a spatial orientation as a result of that force field. We are interested, in particular, in the gravitational field, because it always exists. First we will use thermodynamics to deduce the spatial distribution, and then compute the settling, or the fall, velocity, and finally evaluate the settling rate coefficient.

Spatial Distribution

Consider the change in thermodynamic functions when moving one mole of a gas composed of C_n from zero height to a height z. Then the change in energy ΔE is

$$\Delta E = M_n g z \tag{19}$$

where M_n is the molar mass of C_n and g the acceleration due to gravity. The molar volume can also change, and the increase in entropy ΔS is given by

$$\Delta S = \kappa N \ln(V_z/V_0) \tag{20}$$

where κ is Boltzmann's constant, N Avogadro's number, V_z the molar volume at height z, and V_0 the molar volume at zero height. For a system at equilibrium, $\Delta F = 0$, so that

$$\Delta F = \Delta H - T \Delta S = 0 \tag{21}$$

where F is the Gibbs free energy and H the enthalpy. In our case $\Delta H = \Delta E$, so that equating ΔE and $T \Delta S$ gives

$$V_z = V_0 \exp\{M_n g z / \kappa N T\} \tag{22}$$

Since it is concentration rather than the molar volume that we are interested in, application of the ideal gas law gives

$$[C_n\{z\}] = [C_n\{0\}] \exp\{-M_n g z / \kappa N T\} \tag{23}$$

This is the Boltzmann formula for the concentration distribution in a gravity field.

Settling Velocity

The flux of fall is $[C_n]v_g$, where v_g is the settling, or fall, velocity. The flux of upward motion due to diffusion is $-D_n \partial[C_n]/\partial z$. At equilibrium these two fluxes must be equal

$$[C_n]v_g = -D_n \partial[C_n]/\partial z = D_n M_n g[C_n]/\kappa NT \tag{24}$$

Thus

$$v_g = D_n M_n g/\kappa NT \tag{25}$$

Settling Rate Coefficient

If we consider the situation for which there is no vertical concentration gradient, i.e., for $\partial[C_n]/\partial z = 0$, then particles are settling out of the system, and the rate coefficient k_n for this physical loss process can be calculated directly. The total removal rate is the product of the downward flux and the horizontal cross sectional area so that:

For an infinitely long cylindrical vessel

$$k_n = v_g 2R_0 L/V = 2v_g/\pi R_0 \tag{26a}$$

For a spherical vessel

$$k_n = v_g \pi R_0^2/V = 3v_g/4R_0 \tag{26b}$$

Evaluation of Parameters

In order to evaluate the parameters, we turn to gas kinetic theory. For clarity let us assume that our particles C_n are suspended in a gas, say, N_2. The average velocity of the particles $\bar{v}_n$ is given by

$$\bar{v}_n = (8\kappa T/\pi m_n)^{1/2} \tag{27}$$

where κ is Boltzmann's constant, T the absolute temperature, and m_n the mass of particle C_n. The gas kinetic collision rate coefficient between the particles and the N_2 molecules Z_n is just the product of their average relative velocity and the cross-sectional area of their collision cylinder, i.e.,

$$Z_n = (8\pi\kappa T/\mu_n)^{1/2}(r_{N_2} + r_n)^2 \tag{28}$$

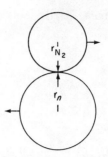

where r_{N_2} and r_n are the collision radii of the N_2 and C_n species, respectively, and μ_n is their reduced mass

$$\mu_n \equiv m_{N_2} m_n / (m_{N_2} + m_n)$$

The mean free path $\bar{l}_n$ is the average velocity divided by the collision frequency $Z_n N_{N_2}$, where N_{N_2} is the number density of N_2 molecules

$$\bar{l}_n = \bar{v}_n / Z_n N_{N_2} = \frac{(\mu_n/m_n)^{1/2}}{\pi N_{N_2} (r_{N_2} + r_n)^2} \qquad (29)$$

For pure N_2, its mean free path $\bar{l}_{N_2}$ is given by

$$\bar{l}_{N_2} = (4\pi \sqrt{2} N_{N_2} r_{N_2}^2)^{-1} \qquad (30)$$

In order to compute the quantities Z_n and $\bar{l}_n$ it is necessary to know r_{N_2}. This quantity is most easily computed from the experimentally measured gas viscosity η_{N_2}, which is related to the other parameters via

$$\eta_{N_2} = \frac{\zeta_\eta}{3} N_{N_2} m_{N_2} \bar{v}_{N_2} \bar{l}_{N_2} \qquad (31)$$

$$= \frac{\zeta_\eta (\kappa T m_{N_2})^{1/2}}{6\pi^{3/2} r_{N_2}^2} \qquad (32)$$

The factor ζ_η is 1.0 in the simple gas kinetic theory. A more rigorous analysis, however gives $\zeta_\eta = 15\pi/32$. (The reader is referred to the work of Hirschfelder et al., 1954 for a more detailed discussion.)

For N_2 at 300°K and 1 atm pressure the parameters of interest are

$$\eta_{N_2} = 177.8 \times 10^{-6} \text{ poise}$$

$$m_{N_2} = 4.6475 \times 10^{-23} \text{ for } {}^{28}N_2$$

$$\bar{v}_{N_2} = 4.7636 \times 10^4 \text{ cm/sec}$$

$N_{N_2} = 2.4474 \times 10^{19}$ molecules/cm³

$r_{N_2} = 1.854 \times 10^{-8}$ cm

$l_{N_2} = 6.6825 \times 10^{-6}$ cm

$Z_{N_2} = 2.9116 \times 10^{-10}$ cm³/molecule-sec

Diffusion Coefficient

The diffusion coefficient D_n for particles C_n diffusing in N_2 can also be obtained from gas kinetic theory

$$D_n\{\text{gas kinetic}\} = \frac{2\zeta_D}{3N_{N_2} m_{N_2}} \frac{(2\kappa T \mu_n)^{1/2}}{\pi^{3/2} (r_{N_2} + r_n)^2} \tag{33}$$

where ζ_D is 1.0 in simple gas kinetic theory, but is $9\pi/16$ from a more detailed analysis (Hirschfelder et al., 1954). The gas kinetic theory derivation applies for a thin gas in which $l_{N_2} \gg r_n$. Actually the denominator of Eq. (33) should contain the total density rather than the diluent gas density $N_{N_2} m_{N_2}$. We assume, however, an infinitely dilute gas.

In the other extreme, i.e., when $l_{N_2} \ll r_n$, the diffusion coefficient can be found from Stokes' law for falling spheres. The terminal velocity is given by

$$v_g = m_n g / 6\pi \eta_{N_2} r_n \tag{34}$$

We have seen, however, that v_g is related to the diffusion coefficient through Eq. (25). Combining Eqs. (25) and (34) gives

$$D_n\{\text{Stokes}\} = \kappa T / 6\pi \eta_{N_2} r_n \tag{35}$$

Equations (33) and (35) do not have the same limiting expressions and cannot be made compatible over the whole range of variables. The gas kinetic expression is the limiting law for very small particles, and the Stokes expression is the limiting law for very large particles. The general expression apparently has never been derived. In the absence of a general expression, the simplest formulation is a linear combination of the two limiting cases. Thus we take D_n as

$$D_n = D_n\{\text{Stokes}\} + D_n\{\text{gas kinetic}\} \tag{36}$$

For a special case of particles with specific gravity 1.0 in N_2 at 1 atm pressure and 300°K, values of the parameters have been computed, and they are listed in Table III. The volumes are equal to the masses since the specific gravity was assumed to be unity. The random "Brownian" average

Table III

Values of Several Parameters for Particles of Specific Gravity 1.0 at 300°K in 1 atm N_2[a]

Particle radius (cm)	Volume (cm^3)	$\bar{v}_n$ (cm/sec)	v_g (cm/sec)	$\bar{l}_n$ (cm)	Z_n (cm^3/sec)	D_n{Stokes} (cm^2/sec)	D_n{gas kin.} / D_n{Stokes}	D_n[b] (cm^2/sec)
3×10^{-8}	1.131 − 22	3.054 + 04	3.503 − 07	2.978 − 06	4.189 − 10	4.119 − 04	3.165 + 02	1.308 − 01
1×10^{-7}	4.189 − 21	5.018 + 03	2.574 − 06	9.695 − 08	2.115 − 09	1.236 − 04	2.090 + 02	2.595 − 02
3×10^{-7}	1.131 − 19	9.657 + 02	9.740 − 06	2.598 − 09	1.519 − 08	4.119 − 05	8.730 + 01	3.637 − 03
1×10^{-6}	4.189 − 18	1.587 + 02	3.612 − 05	4.176 − 11	1.553 − 07	1.236 − 05	2.847 + 01	3.641 − 04
3×10^{-6}	1.131 − 16	3.054 + 01	1.183 − 04	9.150 − 13	1.364 − 06	4.119 − 06	9.724 + 00	4.417 − 05
1×10^{-5}	4.189 − 15	5.018 + 00	4.832 − 04	1.365 − 14	1.502 − 05	1.236 − 06	2.942 + 00	4.872 − 06
3×10^{-5}	1.131 − 13	9.657 − 01	2.188 − 03	2.926 − 16	1.349 − 04	4.119 − 07	9.832 − 01	8.169 − 07
1×10^{-4}	4.189 − 12	1.587 − 01	1.588 − 02	4.331 − 18	1.497 − 03	1.236 − 07	2.952 − 01	1.601 − 07
3×10^{-4}	1.131 − 10	3.054 − 02	1.212 − 01	9.263 − 20	1.347 − 02	4.119 − 08	9.843 − 02	4.524 − 08
1×10^{-3}	4.189 − 09	5.018 − 03	1.262 + 00	1.370 − 21	1.497 − 01	1.236 − 08	2.953 − 02	1.272 − 08
3×10^{-3}	1.131 − 07	9.657 − 04	1.114 + 01	2.929 − 23	1.347 + 00	4.119 − 09	9.844 − 03	4.160 − 09
1×10^{-2}	4.189 − 06	1.587 − 04	1.229 + 02	4.332 − 25	1.497 + 01	1.236 − 09	2.953 − 03	1.239 − 09

[a] *Physical properties of* N_2: viscosity = 177.8×10^{-7} poise at 300°K, mass = 4.6475×10^{-23} g, average velocity = 4.7636×10^4 cm/sec, mean free path = 6.6825×10^{-6} cm, number density = 2.4474×10^{19} per cm^3, molecular diameter = 3.709×10^{-8} cm, collision rate coefficient = 2.9116×10^{-10} cm^3/sec. The entries give the value and the power of 10; e.g., $1.234 − 06 = 1.234 \times 10^{-6}$.

[b] Computed from Eq. (36).

EVALUATION OF PARAMETERS

velocity $\bar{v}_n$ decreases rapidly as the particle radius increases, whereas the gravitational settling velocity v_g increases rapidly. They are equal for particles of about 2×10^{-4} cm radius. Thus larger particles will tend to settle, whereas smaller ones will not.

At 2×10^{-4} cm radius, both $\bar{v}_n$ and v_g are about 7×10^{-2} cm/sec, a rather small velocity. This is the condition where Stokes' law will be more reliable than gas kinetic theory for determining the diffusion coefficient. For either larger or smaller particles, the velocity increases, and fluid dynamics must be considered. The values for D_n and v_g computed at $r_n = 1 \times 10^{-2}$ cm may be considerably in error since v_g is of the order of 100 cm/sec, a nontrivial velocity. For radii smaller than 2×10^{-4} cm, the particle radius is also approaching the mean free path of the gas $\bar{l}_{N_2} = 6.68 \times 10^{-6}$ cm, further compounding the error introduced in Stokes' law. This error is overcompensated for by using Eq. (36) to compute D_n. It can be seen that $D\{\text{Stokes}\} = D\{\text{gas kinetic}\}$ at 3×10^{-5} cm radius, so that a significant correction (actually an over-correction) is made in this region by using Eq. (36) rather than $D_n\{\text{Stokes}\}$. For particles with radii in the region from 10^{-7} to 10^{-4} cm, the values of D_n computed from Eq. (36) tend to be too large, but those computed from Stokes' law are too small.

Equation (36) is extremely convenient for computing diffusion constants, both because it is simple and because the parameters needed for the computation generally are known accurately. But how good is Eq. (36)? A number of determinations of diffusion constants exist, and it has been found that the data fit the empirical formula (Fuchs, 1964):

$$D_n = D_n\{\text{Stokes}\}(1 + b_1 \bar{l}_{N_2}/r_n + (b_2 \bar{l}_{N_2}/r_n) \exp\{-b_3 r_n/\bar{l}_{N_2}\}) \quad (37)$$

where b_1, b_2, and b_3 are empirical parameters for each gas at a given temperature.

Oil droplets in air at atmospheric pressure and 23°C were studied by Millikan (1923) and Mattauch (1925). They fitted their data to Eq. (37), and their b_1, b_2, and b_3 parameters are listed in Table IV. Their data were summarized by Fuchs and his values of the parameters are also listed. A comparison between the values of D_n computed by Eq. (36) to those obtained experimentally [as computed from Eq. (37)] are given in Table IV using the following parameters for air at atmospheric pressure and 23°C:

$$\eta_{\text{air}} = 18.3 \times 10^{-5} \text{ poise}$$

$$N_{\text{air}} = 2.4474 \times 10^{-19} \text{ molecules/cm}^3$$

$$m_{\text{air}} = 4.74 \times 10^{-23} \text{ gm}$$

$$r_{\text{air}} = 1.86 \times 10^{-8} \text{ cm}$$

$$\bar{l}_{\text{air}} = 6.53 \times 10^{-6} \text{ cm}$$

Table IV

Ratios of Calculated to Measured Diffusion Constants[a]

r_n (cm)	Millikan (1923)	Mattauch (1925)	Fuchs (1964)	Reiss (1929)	Knudsen Weber (1911)	Schmitt (1959)
3×10^{-8}	1.53		1.06			
1×10^{-7}	2.84		1.97			
3×10^{-7}	3.51		2.45			
1×10^{-6}	3.67	3.63	2.61	3.89		
3×10^{-6}	3.38	3.44	2.53	3.57		
1×10^{-5}	2.52	2.52	2.13	2.54	2.63	
3×10^{-5}		1.68	1.58	1.68	1.72	1.50
1×10^{-4}		1.23	1.20	1.23	1.24	1.18
3×10^{-4}					1.08	
1×10^{-3}					1.03	
b_1	0.864	0.898	1.246	0.879	0.77	1.45
b_2	0.29	0.312	0.42	0.23	0.40	0.40
b_3	1.25	2.37	0.87	2.61	1.62	0.9

[a] Calculated values from $D\{\text{Stokes}\} + D\{\text{gas kinetics}\}$. Measured values from $D\{\text{Stokes}\} \times [1 + b_1 l_{N_2}/r_n + (b_2 l_{N_2}/r_n) \exp\{-b_3 r_n/l_{N_2}\}]$.

It can be seen that the calculated values are greater than the measured values, by almost a factor of 4 in some cases. With the Fuchs reevaluation, however, the discrepancies are reduced and the largest discrepancy is a factor of 2.61.

Other determinations of diffusion coefficients have been made for aqueous $BaHgI_4$ solution in air (Reiss, 1926) and glass spheres in air (Knudsen and Weber, 1911). The comparison between calculated and measured values shows about the same discrepancy as for oil droplets. A much more recent determination has been made for silicone oil droplets in N_2 (Schmitt, 1959). His data are for a limited range of radii only (7–12 × 10^{-5} cm). The discrepancy between calculated and measured diffusion coefficients is, however, slightly less than for any of the previous determinations.

CHAPTER III

BIMOLECULAR PROCESSES

In this chapter we wish to derive the rate coefficients $k_{m,n}$ for reactions in which two particles coalesce. We shall still assume that the reverse disintegration reaction is negligible, and that the accommodation coefficients are unity.

Rate Coefficient

The gas kinetic theory gives a collision rate coefficient $Z_{n,m}$ between two particles which is the product of their average relative velocity and the cross-sectional area of their collision cylinder:

$$Z_{n,m} = (8\pi\kappa T/\mu_{m,n})^{1/2}(r_n + r_m)^2 \tag{38}$$

where $\mu_{m,n}$ is their reduced mass

$$\mu_{m,n} \equiv m_m m_n/(m_m + m_n)$$

Equation (38) is based on two important assumptions:

1. The two colliding particles are distinguishable, i.e., C_n is not identical to C_m. If the colliding particles are indistinguishable, i.e., for particles of the same chemical composition and size ($n = m$), then $Z_{n,m}$ is $\frac{1}{2}$ the value given by Eq. (38).

2. The collision rate coefficient is based on the concentrations of the species in the region of the collision. In general these concentrations will be smaller than the bulk concentrations because the collisions are depleting the reactants.

An interesting philosophical question is: When are particles gas phase molecules and when do they constitute a separate condensed phase? This question can be answered from a consideration of $Z_{n,m}$. From a kinetics viewpoint, C_n constitutes a solid when its collision frequency with C_m at any temperature depends on the surface area of C_n but on no other property of C_n. An examination of Eq. (38) shows that this will be the case if $n \gg m$, i.e., $r_n \gg r_m$ and $\mu_{n,m} = m_m$. Thus the answer to the question depends on the reaction involved. A particle of 10^{-6}-cm radius constitutes a condensed phase when reacting with particles of radius $< 10^{-7}$ cm. On the other hand particles of 10^{-3}-cm radius are gas phase molecules when reacting with other particles of similar size.

Influence of Diffusion

In general we are interested in the rate coefficient $k_{m,n}$ based on bulk concentrations. The derivation of the rate coefficient taking into account concentration gradients was worked out by Smoluchowski (1916). The derivation follows:

Consider the reactants C_n and C_m separated by some variable distance r. Consider the flux of C_n toward C_m, $-(D_n + D_m)\,\partial[C_n]/\partial r$, where the net diffusion coefficient is given by the sum of the two individual diffusion coefficients. The total flux ψ_m on the whole spherical area is

$$\psi_m = -(D_n + D_m)4\pi r^2 \,\partial[C_n]/\partial r \tag{39}$$

Under steady-state conditions, i.e., for $[C_n]$ independent of time at any distance r from C_m, ψ_m is a constant.

Equation (39) can be separated and integrated as

$$\psi_m \int \partial r/r^2 = -4\pi(D_m + D_n) \int \partial[C_n] \tag{40}$$

This problem is generally solved by considering that when r is infinite, $[C_n]$ is the bulk concentration $[C_n]_\infty$; and that when r is a minimum, i.e.,

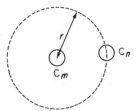

when $r = r_n + r_m$, then $[C_n] = 0$. The last boundary condition is not correct, since C_n must have some finite concentration, say $[C_n]_0$, at $r_n + r_m$. With $[C_n]_0$ as the lower boundary condition for $[C_n]$, the integration gives

$$\psi_m = 4\pi(r_n + r_m)(D_n + D_m)([C_n]_\infty - [C_n]_0) \qquad (41)$$

Since ψ_m is a constant, it can be evaluated at any r. When $r = r_n + r_m$, the flux is given by the gas kinetic collision frequency, i.e.,

$$\psi_m = Z_{n,m}[C_n]_0 \qquad (42)$$

Substitution of Eq. (42) into Eq. (41) leads to

$$[C_n]_0 = [C_n]_\infty/[1 + Z_{n,m}/4\pi(D_n + D_m)(r_n + r_m)] \qquad (43)$$

The rate coefficient $k_{n,m}$ is defined by

$$k_{n,m} \equiv \frac{Z_{n,m}[C_n]_0[C_m]_\infty}{[C_n]_\infty[C_m]_\infty} \qquad (44)$$

Substituting Eq. (43) into Eq. (44) gives

$$k_{n,m} = Z_{n,m}/[1 + Z_{n,m}/4\pi(D_n + D_m)(r_n + r_m)] \qquad (45)$$

Equation (45) is the general expression giving the collision rate coefficient. When diffusion is very fast, i.e., when $Z_{n,m} \ll 4\pi(D_n + D_m)(r_n + r_m)$, then $k_{n,m}$ reduces to the gas kinetic rate coefficient $Z_{n,m}$. When diffusion is very slow, i.e., when $Z_{n,m} \gg 4\pi(D_n + D_m)(r_n + r_m)$, then

$$k_{n,m}\{\text{diffusion controlled}\} = 4\pi(D_n + D_m)(r_n + r_m) \qquad (46)$$

This is the usual expression which was derived by Smoluchowski (1916) with the incorrect boundary condition $[C_n]_0 = 0$.

Values for $Z_{n,m}$ and $k_{n,m}$ (assuming C_n and C_m are distinguishable) are listed in Tables V and VI, respectively, for particles of specific gravity 1.0 at 300°K. $Z_{n,m}$ is independent of the diluent gas or its pressure. The values of $k_{n,m}$ are computed for 1 atm of N_2 as a diluent. Both $Z_{n,m}$ and $k_{n,m}$ increase with particle size, but $Z_{n,m}$ increases more rapidly. For small particles $Z_{n,m}$ is equal to $k_{n,m}$ but for particles of 10^{-6}-cm radius and larger, diffusional effects become significant.

Table V

Values of Gas Kinetic Collision Rate Coefficient $Z_{n,m}$ (cm^3/sec) for Particles of Specific Gravity 1.0 at 300°K[a]

r_m (cm)	3 x 10^{-8}	1 x 10^{-7}	3 x 10^{-7}	1 x 10^{-6}	3 x 10^{-6}
3 x 10^{-8}	4.884 - 10				
1 x 10^{-7}	1.643 - 09	8.917 - 10			
3 x 10^{-7}	1.045 - 08	2.568 - 09	1.545 - 09		
1 x 10^{-6}	1.018 - 07	1.908 - 08	5.196 - 09	2.820 - 09	
3 x 10^{-6}	8.808 - 07	1.515 - 07	3.305 - 08	8.122 - 09	4.884 - 09
1 x 10^{-5}	9.651 - 06	1.608 - 06	3.218 - 07	6.035 - 08	1.643 - 08
3 x 10^{-5}	8.651 - 05	1.428 - 05	2.785 - 06	4.791 - 07	1.045 - 07
1 x 10^{-4}	9.599 - 04	1.580 - 04	3.052 - 05	5.085 - 06	1.018 - 06
3 x 10^{-4}	8.636 - 03	1.420 - 03	2.736 - 04	4.516 - 05	8.808 - 06
1 x 10^{-3}	9.594 - 02	1.577 - 02	3.036 - 03	4.995 - 04	9.651 - 05
3 x 10^{-3}	8.634 - 01	1.419 - 01	2.731 - 02	4.489 - 03	8.651 - 04
1 x 10^{-2}	9.593 + 00	1.576 + 00	3.034 - 01	4.986 - 02	9.599 - 03

[a] $Z_{n,m} = (8\pi\kappa T/\mu_{n,m})^{1/2}(r_n + r_m)^2$ assumes collision partners are distinguishable even when $n = m$. Entries give the value and the power of 10; e.g., $1.234 - 06 = 1.234 \times 10^{-6}$.

Table VI

Values of Bimolecular Rate Coefficients $k_{m,n}$ (cm^3/sec) for Particles of Specific Gravity 1.0 at 300°K in 1 atm of N_2[a]

r_m (cm)	3 x 10^{-8}	1 x 10^{-7}	3 x 10^{-7}	1 x 10^{-6}	3 x 10^{-6}	1 x 10^{-5}
3 x 10^{-8}	4.872 - 10					
1 x 10^{-7}	1.633 - 09	8.857 - 10				
3 x 10^{-7}	1.026 - 08	2.525 - 09	1.502 - 09			
1 x 10^{-6}	9.602 - 08	1.813 - 08	4.813 - 09	2.443 - 09		
3 x 10^{-6}	7.484 - 07	1.318 - 07	2.717 - 08	5.819 - 09	2.818 - 09	
1 x 10^{-5}	6.087 - 06	1.081 - 06	1.913 - 07	2.764 - 08	5.386 - 09	1.921 - 09
3 x 10^{-5}	3.143 - 05	5.818 - 06	9.251 - 07	1.096 - 07	1.583 - 08	2.573 - 09
1 x 10^{-4}	1.404 - 04	2.705 - 05	3.986 - 06	4.238 - 07	5.432 - 08	6.711 - 09
3 x 10^{-4}	4.665 - 04	9.155 - 05	1.307 - 05	1.337 - 06	1.652 - 07	1.892 - 08
1 x 10^{-3}	1.616 - 03	3.195 - 04	4.504 - 05	4.539 - 06	5.537 - 07	6.176 - 08
3 x 10^{-3}	4.903 - 03	9.716 - 04	1.364 - 04	1.369 - 05	1.664 - 06	1.842 - 07
1 x 10^{-2}	1.641 - 02	3.254 - 03	4.564 - 04	4.572 - 05	5.549 - 06	6.127 - 07

[a] $k_{n,m} = Z_{n,m}/[1 + Z_{n,m}/4\pi(D_n + D_m)(r_n + r_m)]$ assumes collision partners are distinguishable even when $n = m$. Entries give the value and the power of 10; e.g., $1.234 - 06 = 1.234 \times 10^{-6}$.

r_n (cm)						
1×10^{-5}	3×10^{-5}	1×10^{-4}	3×10^{-4}	1×10^{-3}	3×10^{-3}	1×10^{-2}
8.917 − 09						
2.568 − 08	1.545 − 08					
1.908 − 07	5.196 − 08	2.820 − 08				
1.515 − 06	3.305 − 07	8.122 − 08	4.884 − 08			
1.608 − 05	3.218 − 06	6.035 − 07	1.643 − 07	8.917 − 08		
1.428 − 04	2.785 − 05	4.791 − 06	1.045 − 06	2.568 − 07	1.545 − 07	
1.580 − 03	3.052 − 04	5.085 − 05	1.018 − 05	1.908 − 06	5.196 − 07	2.820 − 07

r_n (cm)						
	3×10^{-5}	1×10^{-4}	3×10^{-4}	1×10^{-3}	3×10^{-3}	1×10^{-2}
	1.141 − 09					
	1.548 − 09	7.822 − 10				
	3.537 − 09	1.019 − 09	6.729 − 10			
	1.070 − 08	2.379 − 09	9.415 − 10	6.349 − 10		
	3.123 − 08	6.389 − 09	2.045 − 09	8.458 − 10	6.247 − 10	
	1.031 − 07	2.046 − 08	6.013 − 09	1.928 − 09	8.805 − 10	6.216 − 10

Condensation

Consider the case of the vapor species C condensing on C_n under conditions in which physical source, vaporization, coagulation, and disintegration reactions are negligible. Then the rate equations are

$$\frac{d[C_n]}{dt} = [C](k_{1,n-1}[C_{n-1}] - k_{1,n}[C_n]) - k_n[C_n] \qquad (47)$$

Constant [C]

In general, Eq. (47) has a complex solution. Goodrich (1964a) considered this equation in detail for the situation in which $k_n = 0$, i.e., for the pure condensation process, and solved the problem by using a series expansion for $[C_n]$. Initially he considered the case for the conditions

$$[C] = \text{const}$$
$$[C_n\{0\}] = 0 \quad \text{for} \quad n \geq 2$$

where $[C_n\{0\}]$ is the initial concentration of $[C_n]$ at zero time. The solutions of Goodrich as slightly modified by Hudson and Heicklen (1968) are

$$\frac{[C_n\{\xi\}]}{[C]} = \frac{k'_{1,1}}{2k'_{1,n}} \left(\text{erf}\left\{\frac{\tau_n'}{(2\theta_n^2)^{1/2}}\right\} - \text{erf}\left\{\frac{\tau_n' - \xi}{(2\theta_n^2)^{1/2}}\right\} \right) \qquad (48)$$

where

$$k'_{1,n} \equiv k_{1,n}(2000/T)^{1/2}$$

$$\tau_n' \equiv \sum_{j=2}^{n} (1/k'_{1,j})$$

$$\theta_n^2 \equiv \sum_{j=2}^{n} (1/k'_{1,j})^2$$

$$\xi \equiv t[C](T/2000)^{1/2}$$

The factor $(2000/T)$ was introduced by Hudson and Heicklen to remove the temperature dependence of the rate coefficients which they considered to be collision frequency coefficients. They were interested in the experimental region near 2000°K, so that $2000/T$ was a convenient factor to employ. The equation is equally valid if this factor is dropped in $k'_{1,n}$ and ξ. An interesting feature of Eq. (48) is that $[C_n\{\xi\}]/[C]$ is a function of only n and ξ, i.e., all the dependence on $[C]$ or T is included in ξ.

Equation (48) is good for large n as discussed by Goodrich, but is less accurate for small particles since the expansion was carried out with the use of Chebychev–Hermite polynomials. The results can be effected also by the use of Laguerre polynomials. This results in a convergent series good for small n in which there is, to three terms,

$$\frac{[C_n\{\xi\}]}{[C]} = \frac{k'_{1,1}}{k'_{1,n}\Gamma\{(\tau_n')^2/\theta_n^2\}} \int_0^{\tau_n'\xi/\theta_n^2} x^{[(\tau_n')^2/\theta_n^2 - 1]} e^{-x}\, dx \qquad (49)$$

Hudson and Heicklen (1968) extended the results to include diffusional loss. Rather than using Eq. (47), they treated the diffusion terms exactly for the case of an infinite cylinder. Thus the rate equations were, for $n \geq 2$,

$$\frac{\partial [C_n]}{\partial t} = \frac{D_n}{R}\frac{\partial}{\partial R}\left(R\frac{\partial [C_n]}{\partial R}\right) + k_{1,n-1}[C][C_{n-1}] - k_{1,n}[C][C_n] \qquad (50)$$

where R is the distance from the axis of the cylinder. They still considered

$$[C] = \text{const}$$
$$[C_n\{0\}] = 0, \qquad n \geq 2$$

They effected a solution in an infinite series of zero-order Bessel functions $J_0\{\omega_j R\}$

$$[C_n\{R,t\}] = \sum_{j=1}^{\infty} a_{n,j}\{t\} J_0\{\omega_j R\} \qquad (51)$$

The coefficients ω_j were found from the boundary condition

$$J_0\{\omega_j R_0\} = 0$$

where R_0 is the radius of the reaction vessel. The coefficients $a_{n,j}\{t\}$ were found by expanding in Chebychev–Hermite polynomials, which are good for large n. To three terms in the expansion

$$a_{n,j}\{t\} = \frac{\mu_0\{n,j\}}{2}\Lambda_j\left[\text{erf}\left\{\frac{\tau'_{nj}}{(2\Theta_{nj}^2)^{1/2}}\right\} - \text{erf}\left\{\frac{\tau'_{nj} - t}{(2\Theta_{nj}^2)^{1/2}}\right\}\right] \qquad (52)$$

where

$$\mu_i\{n,j\} \equiv \int_0^\infty t^i a_{n,j}\{t\}\, dt$$

$$\tau'_{nj} \equiv \mu_1\{n,j\}/\mu_0\{n,j\}$$

$$\Theta^2_{nj} = \mu_2\{n,j\}/\mu_0\{n,j\} - (\tau'_{nj})^2$$

$$\Lambda_j \equiv 2k_{1,1}[C]^2/\omega_j J_1\{\omega_j\}$$

and the ω_j are the values of $\omega_j R_0$ for which $J_0\{\omega_j R_0\} = 0$.

Equation (52) is too cumbersome to be of any utility. We are, however, really interested only in the radial average value of $[C_n\{R,t\}]$, denoted as $\overline{[C_n\{t\}]}$ or $\overline{[C_n\{\xi\}]}$. By averaging over the cross section, we find

$$\frac{\overline{[C_n\{\xi\}]}}{[C]} = 2k'_{1,1} \sum_j \frac{\mu_0'\{n,j\}}{k'_{1,n}\omega_j^2} \left(\mathrm{erf}\left\{\frac{\phi_{nj}}{(2\Psi^2_{nj})^{1/2}}\right\} - \mathrm{erf}\left\{\frac{\phi_{nj}-\xi}{(2\Psi^2_{nj})^{1/2}}\right\} \right) \quad (53)$$

where

$$\phi_{nj} \equiv [C](T/2000)^{1/2}\tau'_{nj}$$

$$\Psi^2_{nj} \equiv [C]^2(T/2000)\Theta^2_{nj}$$

$$\mu_0'\{n,j\} \equiv \prod_{l=1}^n \frac{k'_{1,l}}{k'_{1,l} + (D'_l\omega_j^2/[C]PR_0^2)T/2000}$$

and

$$D'_l \equiv D_l P(2000/T)^{3/2}$$

The quantity D'_l was used rather than D_l, since if gas kinetic theory is applicable in determining D_l, then D'_l is independent of the pressure P and temperature T.

Since Chebychev–Hermite polynomials were used in the series expansion, Eq. (53) is accurate for large n. For small n, more accurate values are obtained with Laguerre polynomials:

$$\frac{\overline{[C_n\{\xi\}]}}{[C]} = 4k'_{1,1} \sum_l \frac{\mu_0'\{n,l\}}{k'_{1,n}\omega_l^2 \Gamma\{\phi^2_{nl}/\Psi^2_{nl}\}} \int_0^{\phi_{nl}\xi\Psi^2_{nl}} x^{(\phi^2_{nl}/\Psi_n^2 - 1)} e^{-x}\, dx \quad (54)$$

An important feature of Eqs. (53) and (54) is that $\overline{[C_n\{\xi\}]}/[C]$ depends on only three variables if the $k_{1,j}$ and D_j are evaluated from gas kinetic theory. These variables are n, ξ, and Ω, where

$$\Omega \equiv T/[C]PR_0$$

and ξ is still

$$\xi \equiv t[C](T/2000)^{1/2}$$

The values of $\overline{[C_n\{\xi\}]}/[C]$ have been evaluated on a high-speed computer for interesting values of n, ξ, and Ω by Hudson and Heicklen (1968). These results are shown in Fig. 1. The units used for ξ and Ω are, respectively, M sec and $°K\ M^{-1}\ atm^{-1}\ cm^{-2}$.

The general behavior of the function $\overline{[C_n\{\xi\}]}/[C]$ is that for any n, when the time parameter is sufficiently short, $[C_n] = 0$. As ξ reaches some critical value (say 10^{-9} for $n = 10^4$), then $[C_n]$ rises in a step function to its steady-state value, given by the solid lines. These steady-state values rise as Ω decreases, reaching an upper limiting value when $\Omega = 0$, i.e., when the physical loss term due to wall deposition is negligible.

The steady-state values of $[C_n]$ are

$$\frac{\overline{[C_n\{\infty\}]}}{[C]} = [C]^{n-1} \prod_{j=2}^{n} \frac{k_{1,j-1}}{k_{1,j}[C] + k_j} \quad (55)$$

It is interesting to examine the mass distribution among the C_n for steady-state conditions. This was done by Hudson and Heicklen (1968) and their results are shown in Fig. 2. For $\Omega < \sim 3.75 \times 10^{13}\ °K\ M^{-1}\ atm^{-1}\ cm^{-1}$, the mass for any C_n increases with n, but at lower Ω, the reverse is true.

The drawback of these considerations for pure condensation is that they are never applicable in a real situation because vaporization becomes more and more important as the drops decrease in size. A simple variation

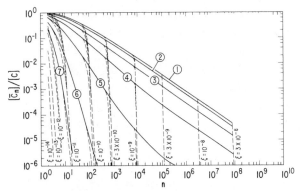

Fig. 1 Log–log plots of $[\overline{C}_n]/[C]$ versus n for various values of Ω and ξ. Curve 1 is for $\Omega = 0$; curve 2 for $\Omega = 3.75 \times 10^{11}$; curve 3 for $\Omega = 3.75 \times 10^{12}$; curve 4 for $\Omega = 1.25 \times 10^{13}$; curve 5 for $\Omega = 3.75 \times 10^{13}$; curve 6 for $\Omega = 1.25 \times 10^{14}$; curve 7 for $\Omega = 3.75 \times 10^{14}$. From Hudson and Heicklen (1968) with permission of Pergamon Press.

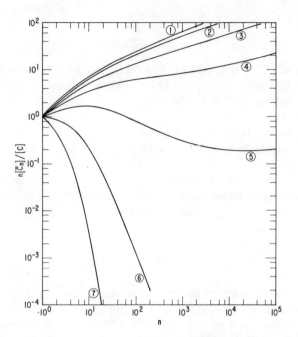

Fig. 2 Log–log plots of $n[\overline{C_n}]/[C]$ versus n for various values of Ω with $\xi = \infty$. Curve 1 is for $\Omega = 0$; curve 2 for $\Omega = 3.75 \times 10^{11}$; curve 3 for $\Omega = 3.75 \times 10^{12}$; curve 4 for $\Omega = 1.25 \times 10^{13}$; curve 5 for $\Omega = 3.75 \times 10^{13}$; curve 6 for $\Omega = 1.25 \times 10^{14}$; curve 7 for $\Omega = 3.75 \times 10^{14}$. From Hudson and Heicklen (1968) with permission of Pergamon Press.

was utilized by de Pena et al. (1973) who found that nucleation occurred in a burst and then no more small particles were formed. The particles already present grew by condensation. Thus de Pena et al. considered the case of particles of sizes only in excess of some critical size q which underwent condensation reactions. The conditions then were

$$[C] = \text{const}$$
$$[C_n] = 0 \quad \text{for all time}, \quad n < q$$
$$[C_q\{0\}] = N \quad \text{at } t = 0$$
$$[C_n\{0\}] = 0 \quad \text{at } t = 0, \quad n > q$$

where N is the total number of particles at zero time.

Equation (47) now has a simple solution if the approximation is made that $k_{1,n-1} = k_{1,n}$; a condition which in fact gets better and better as $n \Rightarrow \infty$. Equation (47) is then the equation of the Poisson process, and the solution is

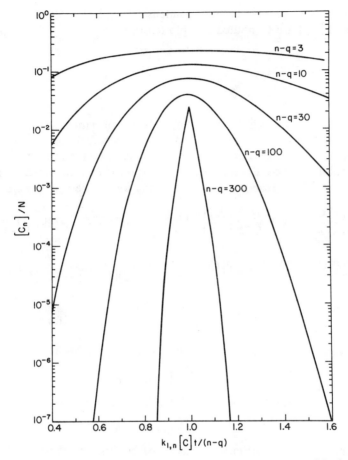

Fig. 3 Semilog plot of $[C_n]/N$ versus $k_{1,n}[C]t/(n-q)$ for various size particles for $[C]$ = const, $[C_n] = 0$ for $1 < n < q$, $[C_q] = N$ at $t = 0$, and $[C_n] = 0$ at $t = 0$ for $n > q$.

$$\frac{[C_n]}{N} = \frac{(k_{1,n}[C]t)^{n-q}}{(n-q)!} \exp\{-(k_{1,n}[C] + k_n)t\} \qquad (56)$$

This solution is strongly peaked in n at any time t and is strongly peaked in t at any n. For any value of n, a maximum in $[C_n]$ occurs when

$$(k_{1,n}[C] + k_n)t = n - q \qquad (57)$$

As the condensation proceeds the concentration of C_q drops exponentially, and larger particles are formed with an ever-decreasing peak value as n

increases. For the case of pure condensation, i.e., for all the $k_n = 0$, the values of $[C_n]$ are computed and plotted in Fig. 3.

Variable [C]

The problem in which [C] is no longer constant was considered by Goodrich (1964a) and extended to include physical loss terms through diffusion by Heicklen et al. (1969). In the more general treatment of Heicklen et al. solutions were developed for a constant chemical source term Q for C in a cylindrical reaction vessel. The equations to be solved were

$$\frac{\partial [C]}{\partial t} = Q + \frac{D_1 \partial}{R \partial R}\left(R \frac{\partial [C]}{\partial R}\right) - \sum_{j=2}^{\infty} k_{1,j}[C][C_j] - 2k_{1,1}[C]^2 \quad (58)$$

$$\frac{\partial [C_n]}{\partial t} = \frac{D_n}{R} \frac{\partial}{\partial R}\left(R \frac{\partial [C_n]}{\partial R}\right) + k_{1,n-1}[C][C_{n-1}] - k_{1,n}[C][C_n]$$

$$\text{for} \quad n > 1. \quad (59)$$

These equations were subjected to the boundary conditions

at $t = 0$, $[C_n] = 0$,

at $R = R_0$, $[C_n] = 0$, $n \geq 1$

at $R = 0$, $\partial [C_n]/\partial R = 0$,

It should be noted that in both this treatment and in the previous one for constant [C], it is assumed that $[C_n] = 0$ at the wall, an assumption which is in fact not correct. The error introduced, however, is only significant near the wall, so that radial averaged concentrations will be sufficiently accurate for all practical purposes.

To solve the equations, Heicklen et al. considered the solutions as infinite series of Bessel functions

$$[C_n] = \sum_{j=1}^{\infty} \frac{\omega_j}{2J_1\{\omega_j\}} B_{nj}\{t\} J_0\{\omega_j R\}, \quad n \geq 2 \quad (60)$$

where J_1 is the Bessel function of order one. The coefficients $B_{nj}\{t\}$ were obtained numerically on a high-speed computer by taking $k_{1,n}$ and D_n from gas kinetic theory (i.e., $k_{1,n} = Z_{1,n}$). The interesting feature of the

CONDENSATION

radial averaged solution for $[C_n]$ is that although there are six explicit variable parameters (n, t, Q, R_0, P, and T), these could be combined so that the solutions depended on n and only three other variables

$$\xi \equiv t\left(\frac{A_0}{8.1 \times 10^{-6}\ M}\right)\left(\frac{T}{2000°K}\right)^{1/2} \text{ sec}$$

$$\chi = T/PA_0R_0^2 \text{ °K/atm } M \text{ cm}^2$$

$$\theta = k_0/T^{1/2} \quad M^{-1} \text{ sec}^{-1} \text{ °K}^{-1/2}$$

where it was assumed that the chemical production term Q was a second-order reaction of the form

$$Q = k_0 A_0^2 \tag{61}$$

where k_0 is the second-order rate coefficient and A_0 is the initial pressure of the source reactant. The time parameter ξ is modified from that used above in that $[C]$ is now replaced by A_0. The 2000°K and $8.1 \times 10^{-6}\ M$ were normalizing factors used for convenience.

The numerical results for $\theta = 722\ M^{-1} \text{ sec}^{-1} \text{ °K}^{-1/2}$ are shown graph-

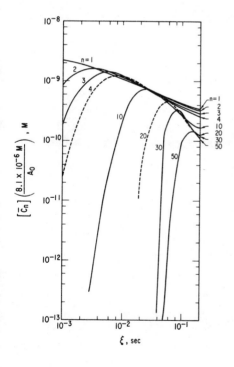

Fig. 4 Log–log plots of the concentration parameter versus the time parameter for various size species at $\theta = 722\ M^{-1} \text{ sec}^{-1} \text{ °K}^{-1/2}$ for $\chi = 4.64 \times 10^8$ °K M^{-1} atm^{-1} cm^{-2}. From Heicklen et al. (1969) with permission of Pergamon Press.

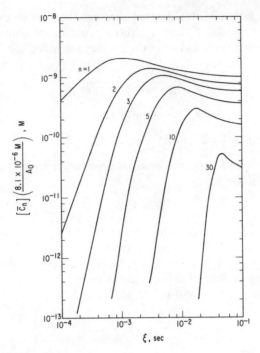

Fig. 5 Log–log plots of the concentration parameter versus the time parameter for various size species at $\theta = 722\ M^{-1}\ \text{sec}^{-1}\ °K^{-1/2}$ for $\chi = 4.64 \times 10^9\ °K\ M^{-1}\ \text{atm}^{-1}\ \text{cm}^{-2}$. From Heicklen et al. (1969) with permission of Pergamon Press.

ically in Figs. 4–6 for three values of χ. As χ increases, diffusional loss becomes more important. The final concentration of C increases, however, because the diffusional loss of the higher C_n reduces their concentration, and they do not remove C as rapidly.

Pure Coagulation

The total coagulation rate is given by

$$\frac{-d\text{N}}{dt} = \sum_{n \geq m} k_{n,m}[C_n][C_m] \tag{62}$$

where $\text{N} = \sum_n [C_n]$, i.e., the particle number density, and n and m are integers which are large enough so that C_n and C_m are considered particles (for total coalescence, n and m include all the positive integers). The summation extends over all applicable values of n and m. The rate coefficients

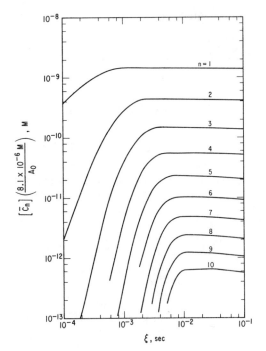

Fig. 6 Log–log plots of the concentration parameter versus the time parameter for various size species at $\theta = 722\ M^{-1}\ \text{sec}^{-1}\ °K^{-1/2}$ for $\chi = 4.64 \times 10^{10}\ °K\ M^{-1}\ \text{atm}^{-1}\ \text{cm}^{-2}$. From Heicklen et al. (1969) with permission of Pergamon Press.

$k_{n,m}$ are those for distinguishable particles, as computed from Eq. (45), except when $n = m$. Then the particles are indistinguishable and one-half the values computed from Eq. (45) must be used.

Usually it is convenient to express the coagulation rate in terms of some average rate coefficient k_{coag}

$$-dN/dt = k_{coag} N^2 \qquad (63)$$

Thus k_{coag} is defined by

$$k_{coag} \equiv \frac{1}{N^2} \sum_{n \geq m} k_{m,n}[C_n][C_m]$$

A question of considerable importance is what becomes of system of coagulating particles? Does it reach some final self-preserving shape? This question has been examined in detail by Friedlander and his co-workers (Friedlander and Wang, 1966; Wang and Friedlander, 1967; Pich et al., 1970).

The equations to be treated are

$$\partial[C_n]/\partial t = \sum_{j=1}^{n-1} k_{j,n-j}[C_j][C_{n-j}] - \sum_{j=1}^{\infty} k_{n,j}[C_n][C_j] - k_{n,n}[C_n]$$

for $n > 1$ (64)

$$\partial[C]/\partial t = -\sum_{j=1}^{\infty} k_{1,j}[C][C_j] - k_{1,1}[C]^2 \qquad \text{for} \quad n = 1 \quad (65)$$

Pich *et al.* also considered the cases where production terms are included in Eq. (65) or where [C] could be maintained at some fixed level. The treatment of the exact equations does not lead to analytical solutions, so many approximations are needed. As a result of their and other work, however,

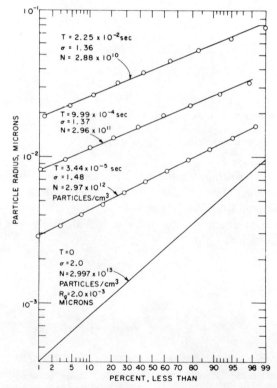

Fig. 7 Size distribution for coagulating particles with an initial geometric standard deviation of 2.0. From Lindauer and Castleman (1971a) with permission of The American Nuclear Society.

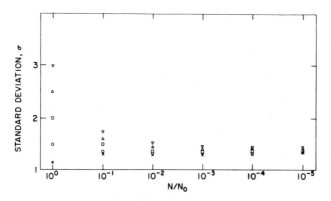

Fig. 8 Approach of initial log–normal distributions to a self-preserving distribution. Values given for σ_i are 1.15 ($\times$), 1.5 ($\bigcirc$), 2.0 ($\square$), 2.5 ($\triangle$), and 3.0 ($\triangledown$). From Lindauer and Castleman (1971a) with permission of The American Nuclear Society.

it is generally agreed that a collection of coagulating particles approaches a log–normal distribution

$$n\{r\} = \frac{N}{(2\pi)^{1/2} r \ln \sigma} \exp\left\{ -\frac{\ln(r/r_g)}{\sqrt{2} \ln \sigma} \right\} \tag{66}$$

where $n\{r\}$ is the concentration of particles of radius r, r_g the geometric

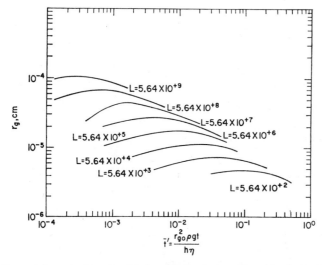

Fig. 9 Decrease in aerosol number density with time for particles with a basis of $r_{go} = 0.01\ \mu$ and $\sigma_0 = 1.37$. (See List of Symbols for parameter definitions.) From Lindauer and Castleman (1971b) with permission of Pergamon Press.

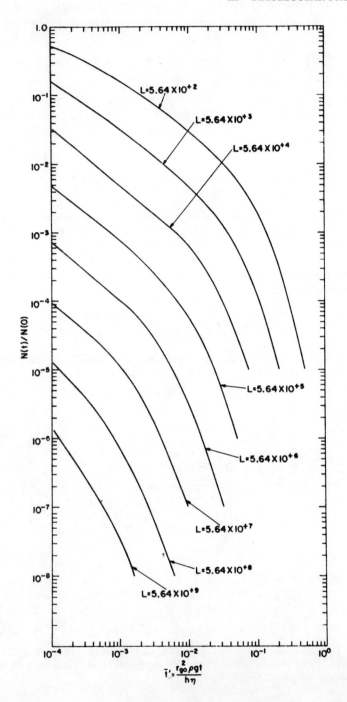

mean radius, and σ the geometric standard derivation. Such a distribution is found in natural and atmospheric aerosols.

Lindauer and Castleman (1971a) performed computer calculations for a system of coagulating particles initially consisting of $>10^9$ particles/cm^3 in a log–normal distribution to determine their time history. Starting with $\sigma = 2.0$, they found that the log–normal distribution was preserved with σ decreasing to 1.36 at infinite time. Their results are shown in Fig. 7. In fact, regardless of the starting value of σ, the log–normal distribution was maintained with σ approaching a value between 1.34 and 1.44, the lower value being for their lowest starting σ of 1.15 and the upper value being for their highest starting σ of 3.0. These results are summarized in Fig. 8. In this work they used an average coagulation constant k_{coag} in place of $k_{n,m}$.

Lindauer and Castleman (1971b) extended their work to include gravitational settling. With $\sigma = 1.37$, the particle number density as a function of time is shown in Fig. 9. For various conditions in this figure the symbols refer to h the fall height, η the gas viscosity, ρ the density, g the acceleration due to gravity, r_{go} the initial geometric mean radius, and $N\{0\}$ the initial number density. The particle number density falls off with time, the fall-off being more rapid the larger the value of the parameter L, defined as

$$L \equiv N\{0\}k_{coag}h\eta/\rho g r_{go}^2$$

At later times, when sufficient large particles are present, gravitational settling becomes important, and the fall-off is enhanced. The geometric mean radius first increases as the particles grow, but then decreases as gravitational settling removes the larger particles as shown in Fig. 10. Thus the gravitational settling also disrupts the log–normal distribution for the larger particles.

Fig. 10 Variation of geometric mean radius with time. The parameters used are defined in the text discussion with a basis of $r_{go} = 0.01$ μ and $\sigma_0 = 1.37$. From Lindauer and Castleman (1971b) with permission of Pergamon Press.

CHAPTER IV

THERMODYNAMICS AND REVERSE REACTIONS

In the previous chapters we have derived rate coefficients for diffusional loss and coalescence processes. We now turn our attention to the reverse reactions—physical production sources (e.g., wall vaporization) and particle disintegration, respectively. We approach these processes through thermodynamics, since at equilibrium the ratio of forward and reverse rate coefficients equals the equilibrium constant.

Vaporization

Equilibrium Constant

We consider first the reactions involving C

$$C + C_n \rightleftarrows C_{n+1} \tag{67}$$

VAPORIZATION

with a forward rate coefficient $k_{1,n}$, a reverse rate coefficient $k_{-1,n}$, and an equilibrium constant $K_{1,n}$, such that, at equilibrium

$$K_{1,n} = k_{1,n}/k_{-1,n} \tag{68}$$

The equilibrium constant is related to the change in Gibbs free energy at standard states $\Delta F^0_{1,n}$, through the basic equation

$$\Delta F^0_{1,n} = -\kappa T \ln K_{1,n} \tag{69}$$

Let us assume that $\Delta F^0_{1,n}$ is identical for all n, except for the surface energy terms, i.e., the molar volume free energy is independent of the volume. Of course this may break down for n very small, but we circumvent this possibility by defining the surface energy in such a way as to include all deviations from the macroscopic volume free energy. For the time being, however, let us simplify the situation and say that the surface tension γ is constant for particles of any size, i.e., for all C_n, for $n > 1$. For $n = 1$, i.e., the fundamental vapor species C, there is no surface energy.

The increase in surface energy, $\Delta F^{\text{surf}}_{1,n}$, in reaction (67) is then due to the surface expansion when $C_n \to C_{n+1}$:

$$\Delta F^{\text{surf}}_{1,n} = \begin{cases} \gamma 4\pi(r^2_{n+1} - r_n^2), & n > 1 \tag{70} \\ \gamma 4\pi r_2^2, & n = 1 \tag{71} \end{cases}$$

where r_n is the equivalent spherical radius of C_n. It is proportional to $n^{1/3}$

$$r_n = r_C n^{1/3} \tag{72}$$

where r_C is the *apparent* molecular radius as deduced from the macroscopic density. Then

$$\Delta F^{\text{surf}}_{1,n} = \begin{cases} \gamma 4\pi r_C^2 [(n+1)^{2/3} - n^{2/3}], & n > 1 \tag{73} \\ \gamma 4\pi r_C^2 (2^{2/3}), & n = 1 \tag{74} \end{cases}$$

When n is very large, $(n+1)^{2/3}$ can be expanded by the binomial theorem to give $n^{2/3}(1 + 2/3n + \cdots)$. Thus for $n \Rightarrow \infty$, $\Delta F^{\text{surf}}_{1,n} \Rightarrow 0$.

We can deduce $K_{1,n}$ since

$$\Delta F^0_{1,n} - \Delta F^0_{1,\infty} = \Delta F^{\text{surf}}_{1,n} \tag{75}$$

$$-\kappa T \ln(K_{1,n}/K_{1,\infty}) = \begin{cases} 4\pi r_C^2 \gamma [(n+1)^{2/3} - n^{2/3}], & n > 1 \tag{76} \\ 4\pi r_C^2 \gamma (2^{2/3}), & n = 1 \tag{77} \end{cases}$$

Now $K_{1,\infty}$ is the equilibrium constant between the vapor and an infinitely large surface. It is usually taken as the reciprocal equilibrium vapor pressure of C, $[C]_{vp}^{-1}$, at any temperature. Thus we have the general expression for the equilibrium constant

$$\ln([C]_{vp} K_{1,n}) = \frac{-4\pi r_C^2 \gamma}{\kappa T} [(n+1)^{2/3} - n^{2/3}], \quad n > 1 \quad (78)$$

$$= -4\pi r_C^2 \gamma (2^{2/3})/\kappa T, \quad n = 1 \quad (79)$$

Rate Coefficient

At equilibrium, i.e., when $k_{1,n} = Z_{1,n}$, then

$$k_{-1,n} = k_{1,n}/K_{1,n} = Z_{1,n}/K_{1,n} \quad (80)$$

so that

$$k_{-1,n} = Z_{1,n}[C]_{vp} \exp\left\{\frac{4\pi r_C^2 \gamma}{\kappa T}[(n+1)^{2/3} - n^{2/3}]\right\}, \quad n > 1 \quad (81)$$

$$= Z_{1,1}[C]_{vp} \exp\{4\pi r_C^2 \gamma (2^{2/3})/\kappa T\}, \quad n = 1 \quad (82)$$

Near any particle C_{n+1}, the rate of condensation is $Z_{1,n+1}[C]$, and the rate of vaporization is $k_{-1,n}$. The net rate of deposition, i.e., the rate of condensation minus the rate of vaporization, is conveniently discussed in terms of an effective collision rate coefficient $Z_{1,n+1}\{\text{eff}\}$, defined by

$$\text{net rate of deposition on } C_{n+1} \equiv Z_{1,n+1}\{\text{eff}\}[C]_0 \quad (83)$$

where $[C]_0$ is the concentration of $[C]$ near C_{n+1}. Thus

$$Z_{1,n+1}\{\text{eff}\} = Z_{1,n+1}\left(1 - \frac{Z_{1,n}}{Z_{1,n+1}}\frac{[C]_{vp}}{[C]_0}\right.$$
$$\left. \times \exp\left\{\frac{4\pi r_C^2 \gamma}{\kappa T}[(n+1)^{2/3} - n^{2/3}]\right\}\right),$$
$$n > 1 \quad (84)$$

$$= Z_{1,1}\left(1 - \frac{Z_{1,1}}{Z_{1,2}}\frac{[C]_{vp}}{[C]_0}\exp\{4\pi r_C^2 \gamma (2^{2/3})/\kappa T\}\right),$$
$$n = 1 \quad (85)$$

These general expressions take on simpler forms if $n \gg 1$. In that case $Z_{1,n}/Z_{1,n+1} \Rightarrow 1$, and $(n+1)^{2/3} \Rightarrow n^{2/3}(1 - 2/3n)$. The expression for $Z_{1,n+1}\{\text{eff}\}$ simplifies to

$$Z_{1,n+1}\{\text{eff}\} = Z_{1,n+1}\left(1 - \frac{[C]_{vp}}{[C]_0}\exp\left\{\frac{8\pi r_C^2 \gamma}{3\kappa T n^{1/3}}\right\}\right) \quad (86)$$

If n becomes infinite, i.e., for a plane surface, then $Z_{1,n+1}$ is replaced by the wall collision speed $k_1\{\text{wall}\} = \bar{v}_1/4$, and the net effective wall rate removal speed $k_1\{\text{eff wall}\}$, becomes

$$k_1\{\text{eff wall}\} = \frac{\bar{v}_1}{4}\left(1 - \frac{[C]_{\text{vp}}}{[C]_0}\right) \tag{87}$$

It should be noticed that any of the effective condensation coefficients can become negative. This occurs when vaporization is faster than condensation.

Spontaneous Fracture

Equilibrium Constant

Let us consider the class of reactions

$$C_2 + C_n \rightleftarrows C_{n+2} \tag{88}$$

with forward and reverse rate coefficients, respectively, of $k_{2,n}$ and $k_{-2,n}$ which, under equilibrium conditions, are related to the equilibrium constant $K_{2,n}$ by

$$K_{2,n} = k_{2,n}/k_{-2,n} \tag{89}$$

This equilibrium can be considered in terms of three other reactions, of which we have more detailed information, viz.,

$$C + C_n \rightleftarrows C_{n+1}, \quad K_{1,n}$$
$$C + C_{n+1} \rightleftarrows C_{n+2}, \quad K_{1,n+1}$$
$$2C \rightleftarrows C_2, \quad K_{1,1}$$

From the preceding section we can compute these three equilibrium constants from macroscopic properties (i.e., γ and r_C). Reaction (88) in which we are interested is just the sum of the first two of these reactions minus the third. Thus the equilibrium constants are related by

$$K_{2,n} = K_{1,n}K_{1,n+1}/K_{1,1} \tag{90}$$

from which

$$\ln([C]_{\text{vp}}K_{2,n}) = -\frac{4\pi r_C^2 \gamma}{\kappa T}[(n+2)^{2/3} - n^{2/3} - 2^{2/3}], \quad n > 1 \tag{91}$$

In a similar manner we can consider the next class of reactions

$$C_3 + C_n \rightleftarrows C_{n+3} \tag{92}$$

with rate coefficients $k_{3,n}$ and $k_{-3,n}$ and an equilibrium constant $K_{3,n}$. This reaction is related to ones whose equilibrium constants we already know

$$C_2 + C_n \rightleftarrows C_{n+2}, \quad K_{2,n}$$
$$C + C_{n+2} \rightleftarrows C_{n+3}, \quad K_{1,n+2}$$
$$C + C_2 \rightleftarrows C_3, \quad K_{1,2}$$

In the general case

$$K_{m+1,n} = K_{m,n} K_{1,m+n}/K_{1,m} \tag{93}$$

Thus

$$\ln([C]_{vp} K_{m+1,n}) = \frac{-4\pi r_C^2 \gamma}{\kappa T}[(m+1+n)^{2/3} - (m+1)^{2/3} - n^{2/3}],$$
$$m,n > 1 \quad (94)$$

Rate Coefficient

At equilibrium, i.e., when $k_{m,n} = Z_{m,n}$, then

$$k_{-m,n} = k_{m,n}/K_{m,n} = Z_{m,n}/K_{m,n}$$

$$k_{-m,n} = Z_{m,n}[C]_{vp} \exp\left\{\frac{4\pi r_C^2 \gamma}{\kappa T}[(n+m)^{2/3} - m^{2/3} - n^{2/3}]\right\},$$
$$m,n > 1 \quad (95)$$

Consider the case in which many particles C_m are in the vicinity of a larger particle C_{n+m}. Then the rate of deposition of C_m on C_{n+m} is $Z_{m,m+n}[C_m]_0$ and the rate of loss of C_m from C_{n+m} is $k_{-m,n}$. The net rate of deposition in this process, i.e., the rate of coagulation minus the rate of spontaneous fracture of C_{n+m} into $C_n + C_m$, is conveniently discussed, as with condensation, in terms of an effective collision rate coefficient $Z_{m,n+m}\{\text{eff}\}$ defined by

$$\text{net rate of deposition on } C_{n+m} \equiv Z_{m,n+m}\{\text{eff}\}[C_m]_0 \tag{96}$$

Thus

$$Z_{m,n+m}\{\text{eff}\}$$
$$= Z_{m,n+m}\left(1 - \frac{Z_{m,n}}{Z_{m,n+m}} \frac{[C]_{vp}}{[C_m]_0} \exp\left\{\frac{4\pi r_C^2 \gamma}{\kappa T}[(m+n)^{2/3} - m^{2/3} - n^{2/3}]\right\}\right)$$
$$m,n > 1 \quad (97)$$

If $n \gg m$ the preceding general expression simplifies to

$$Z_{m,n+m}\{\text{eff}\} = Z_{m,n+m}\left(1 - \frac{[C]_{vp}}{[C_m]_0}\frac{Z_{m,n}}{Z_{m,n+m}}\right.$$
$$\left. \times \exp\left\{\frac{4\pi r_C^2 \gamma}{\kappa T}\left[\frac{2m}{3n^{1/3}} - m^{2/3}\right]\right\}\right), \quad m > 1 \quad (98)$$

If n becomes infinite, i.e., at a plane surface, then $Z_{n,m+n}$ is replaced by the wall collision speed $k_m\{\text{wall}\} = \bar{v}_m/4$, and the effective wall rate removal speed $k_m\{\text{eff wall}\}$, becomes

$$k_m\{\text{eff wall}\} = \frac{\bar{v}_m}{4}\left(1 - \frac{[C]_{vp}}{[C_m]}\exp\left\{\frac{-4\pi r_C^2 \gamma}{\kappa T}m^{2/3}\right\}\right), \quad m > 1 \quad (99)$$

Supersaturation

The supersaturation S_m of each species C_m is defined as the ratio of its concentration to the equilibrium vapor pressure of C

$$S_m \equiv [C_m]/[C]_{vp}$$

When particles of uniform size C_m are near a larger particle C_{n+m}, there is some concentration of C_m such that the rate of coalescence equals the rate of disintegration of C_{n+m} to give C_m, i.e., $Z_{n,n+m}\{\text{eff}\} = 0$. At this concentration there is a steady state, and S_m is at the steady-state supersaturation $S_{m,n+m}$. The expression for $S_{m,n+m}$ can be deduced from Eqs. (84), (85), and (97) to be

$$S_{m,n+m} = \frac{(m+n)}{[n(n+2m)]^{1/2}}\left[\frac{m^{1/3}+n^{1/3}}{m^{1/3}+(m+n)^{1/3}}\right]^2$$
$$\times \exp\left\{\frac{4\pi r_C^2 \gamma}{\kappa T}[(n+m)^{2/3} - n^{2/3} - m^{2/3}]\right\}, \quad n,m > 1 \quad (100)$$

$$S_{1,n+1} = \frac{(1+n)}{[n(n+2)]^{1/2}}\left[\frac{1+n^{1/3}}{1+(n+1)^{1/3}}\right]^2$$
$$\times \exp\left\{\frac{4\pi r_C^2 \gamma}{\kappa T}[(1+n)^{2/3} - n^{2/3}]\right\}, \quad n > 1 \quad (101)$$

$$S_{1,2} = \frac{2}{3^{1/2}}\left[\frac{2}{1+2^{1/3}}\right]^2 \exp\left\{\frac{4\pi r_C^2 \gamma}{\kappa T}2^{2/3}\right\} \quad (102)$$

Table VII gives values of $S_{m,n+m}$ for various m and n for H_2O droplets at 25°C in air. Under these conditions the density of liquid H_2O is 0.997044 gm/cc and

$$\gamma = 71.97 \quad \text{dyn/cm}$$
$$r_C = 1.92 \times 10^{-8} \quad \text{cm}$$
$$4\pi r_C^2 \gamma / \kappa T = 8.1595$$

It should be emphasized that there is a fundamental difference between C and all the other C_m with $m > 1$. This is because C has no surface tension associated with it. Thus condensation–vaporization (removal and formation of C) behaves differently than coagulation–spontaneous fracture (removal and formation of C_m for $m > 1$). This is reflected in the values of $S_{m,n+m}$ in Table VII. For $S_{1,n+1}$ and $S_{m,m+1}$ the values are always >1,

Table VII

Values of $S_{m,n+m}$ for $4\pi r_C^2 \gamma / \kappa T = 8.1595$[a]

n	1	2	3	4	6	8	10	12
1	3.815 + 05	5.227 + 01	3.553 + 01	2.776 + 01	2.045 + 01	1.697 + 01	1.495 + 01	1.362 +
2	5.059 + 01	4.319 − 03	2.131 − 03	1.314 − 03	6.932 − 04	4.567 − 04	3.382 − 04	2.688 −
3	3.328 + 01	2.095 − 03	8.224 − 04	4.239 − 04	1.710 − 04	9.274 − 05	5.910 − 05	4.163 −
4	2.517 + 01	1.272 − 03	4.190 − 04	1.871 − 04	6.039 − 05	2.769 − 05	1.546 − 05	9.766 −
6	1.745 + 01	6.493 − 04	1.655 − 04	5.939 − 05	1.345 − 05	4.670 − 06	2.079 − 06	1.087 −
8	1.373 + 01	4.141 − 04	8.782 − 05	2.680 − 05	4.616 − 06	1.282 − 06	4.735 − 07	2.111 −
10	1.153 + 01	2.972 − 04	5.475 − 05	1.472 − 05	2.034 − 06	4.693 − 07	1.479 − 07	5.743 −
12	1.007 + 01	2.294 − 04	3.774 − 05	9.147 − 06	1.052 − 06	2.076 − 07	5.702 − 08	1.963 −
14	9.030 + 00	1.858 − 04	2.784 − 05	6.188 − 06	6.093 − 07	1.049 − 07	2.556 − 08	7.902 −
16	8.245 + 00	1.558 − 04	2.157 − 05	4.450 − 06	3.830 − 07	5.851 − 08	1.282 − 08	3.600 −
18	7.630 + 00	1.341 − 04	1.732 − 05	3.351 − 06	2.562 − 07	3.522 − 08	7.011 − 09	1.806 −
20	7.135 + 00	1.176 − 04	1.431 − 05	2.616 − 06	1.800 − 07	2.250 − 08	4.110 − 09	9.783 −
25	6.230 + 00	9.026 − 05	9.708 − 06	1.577 − 06	8.719 − 08	8.918 − 09	1.355 − 09	2.720 −
30	5.613 + 00	7.355 − 05	7.183 − 06	1.064 − 06	4.938 − 08	4.295 − 09	5.611 − 10	9.788 −
40	4.814 + 00	5.436 − 05	4.597 − 06	5.922 − 07	2.107 − 08	1.428 − 09	1.474 − 10	2.060 −
70	3.703 + 00	3.235 − 05	2.128 − 06	2.144 − 07	4.742 − 09	2.038 − 10	1.360 − 11	1.250 −
100	3.205 + 00	2.427 − 05	1.387 − 06	1.216 − 07	2.050 − 09	6.767 − 11	3.493 − 12	2.501 −
300	2.248 + 00	1.197 − 05	4.826 − 07	2.993 − 08	2.542 − 10	4.272 − 12	1.134 − 13	4.218 −
1000	1.722 + 00	7.026 − 06	2.171 − 07	1.033 − 08	5.174 − 11	5.141 − 13	8.094 − 15	1.789 −
3000	1.458 + 00	5.040 − 06	1.318 − 07	5.313 − 09	1.911 − 11	1.364 − 13	1.543 − 15	2.452 −
10^4	1.287 + 00	3.927 − 06	9.075 − 08	3.229 − 09	9.056 − 12	5.040 − 14	4.447 − 16	5.514 −
3×10^4	1.191 + 00	3.364 − 06	7.195 − 08	2.370 − 09	5.693 − 12	2.714 − 14	2.052 − 16	2.180 −
10^5	1.124 + 00	2.997 − 06	6.049 − 08	1.880 − 09	4.024 − 12	1.709 − 14	1.151 − 16	1.089 −
3×10^5	1.085 + 00	2.789 − 06	5.431 − 08	1.629 − 09	3.243 − 12	1.282 − 14	8.033 − 17	7.074 −
∞	1.000 + 00	2.371 − 06	4.256 − 08	1.177 − 09	1.992 − 12	6.692 − 15	3.565 − 17	2.668 −

[a] Entries give the value and the power of 10; e.g., $1.234 - 06 = 1.234 \times 10^{-6}$. $\gamma_n = \gamma_C$ for n ; $\gamma = 0$ for $n = 1$.

whereas for $S_{m,n+m}$ ($n,m > 1$), the values are always <1. It can also be seen from the table that for all species, $S_{m,n+m}$ decreases as either m or n is increased.

Let us examine the case when C and C_2 are both in a steady state with respect to C_n. Then $[C] = [C]_{vp}S_{1,n}$ and $[C_2] = [C]_{vp}S_{2,n}$. Consider the following function

$$S_{2,n}/S_{1,n}^2 = [C]_{vp}[C_2]/[C]^2 \tag{103}$$

Now if C and C_2 are also in equilibrium with each other, then the expression should be a constant equal to $[C]_{vp}K_{1,2}$ for all values of n. An examination of the defining equations (100) and (101) for $S_{2,n}$ and $S_{1,n}$ shows, however, that their ratio depends on n. Thus except for a fortuitous matching at some particular value of n, C and C_2 are, in general, not in equilibrium

m									
14	16	18	20	25	30	40	70	100	
.268 + 01	1.198 + 01	1.145 + 01	1.102 + 01	1.027 + 01	9.780 + 00	9.196 + 00	8.572 + 00	8.449 + 00	
.239 - 04	1.927 - 04	1.698 - 04	1.525 - 04	1.232 - 04	1.051 - 04	8.399 - 05	5.917 - 05	4.995 - 05	
.136 - 05	2.479 - 05	2.030 - 05	1.709 - 05	1.210 - 05	9.310 - 06	6.377 - 06	3.437 - 06	2.498 - 06	
.716 - 06	4.908 - 06	3.755 - 06	2.976 - 06	1.861 - 06	1.298 - 06	7.676 - 07	3.192 - 07	2.005 - 07	
.364 - 07	4.044 - 07	2.736 - 07	1.943 - 07	9.669 - 08	5.619 - 08	2.516 - 08	6.368 - 09	3.013 - 09	
.075 - 07	6.046 - 08	3.668 - 08	2.363 - 08	9.561 - 09	4.699 - 09	1.625 - 09	2.566 - 10	9.239 - 11	
.591 - 08	1.307 - 08	7.198 - 09	4.247 - 09	1.423 - 09	5.990 - 10	1.626 - 10	1.639 - 11	4.528 - 12	
949 - 09	3.642 - 09	1.836 - 09	1.000 - 09	2.818 - 10	1.028 - 10	2.223 - 11	1.458 - 12	3.118 - 13	
907 - 09	1.220 - 09	5.681 - 10	2.876 - 10	6.908 - 11	2.204 - 11	3.848 - 12	1.673 - 13	2.789 - 14	
214 - 09	4.706 - 10	2.037 - 10	9.641 - 11	1.997 - 11	5.623 - 12	8.019 - 13	2.349 - 14	3.076 - 15	
625 - 10	2.028 - 10	8.209 - 11	3.651 - 11	6.590 - 12	1.649 - 12	1.942 - 13	3.895 - 15	4.031 - 16	
834 - 10	9.557 - 11	3.636 - 11	1.526 - 11	2.422 - 12	5.427 - 13	5.328 - 14	7.422 - 16	6.110 - 17	
732 - 11	1.960 - 11	6.503 - 12	2.401 - 12	2.855 - 13	4.979 - 14	3.220 - 15	1.910 - 17	9.105 - 19	
123 - 11	5.464 - 12	1.613 - 12	5.337 - 13	4.944 - 14	6.919 - 15	3.084 - 16	8.364 - 19	2.388 - 20	
621 - 12	7.636 - 13	1.866 - 13	5.157 - 14	3.157 - 15	3.049 - 16	7.172 - 18	4.855 - 21		
470 - 13	2.105 - 14	3.550 - 15	6.838 - 16	1.800 - 17	8.007 - 19	4.737 - 21			
306 - 14	2.606 - 15	3.484 - 16	5.369 - 17	8.255 - 19	2.212 - 20				
040 - 16	1.220 - 17	8.715 - 19	7.240 - 20	2.465 - 22					
157 - 18	1.844 - 19	7.891 - 21							
084 - 19	1.308 - 20								
920 - 20									
021 - 20									
344 - 20									

with each other even though they are each in a steady state with respect to vaporization from and condensation on C_n. By similar reasoning the argument can be extended to any $S_{m,n}$ and $S_{l,n}$. In fact what happens is that if both C and C_m are present, the kinetics leads to the removal of small droplets below some critical size C_q, and the growth of large droplets larger than C_q. The value of q is such that the $[C] = [C]_{vp} S_{1,q}$, and thus q decreases as $[C]$ increases. The final stable thermodynamic situation is that of C in equilibrium with C_∞. This is the general behavior of a condensed phase in equilibrium with its vapor.

At first appearance the preceding phenomenon appears to contradict the law of detailed balancing. This law states that if three components, A, B, and C, coexist, and that A is in equilibrium with B, and B is in equilibrium with C, then A must be in equilibrium with C. In the previous example, however, neither C nor C_2 were in equilibrium with C_n; they were only in a steady state. Consider the situation between C and C_n:

$$C + C_n \rightleftarrows C_{n+1}, \qquad k_{1,n}, k_{-1,n}$$
$$C + C_{n-1} \rightleftarrows C_n, \qquad k_{1,n-1}, k_{-1,n-1}$$

The steady state was such that the rate of the forward reaction $(1,n)$ was equal to the rate of the reverse reaction $(-1, n-1)$. These are not the forward and reverse rates of the same reaction, and thus their equality does not constitute an equilibrium condition.

Results at Low n

The preceding derivations, as well as the computations for $S_{m,n+m}$ in Table VII, are based on the assumption that both γ and r_C, obtained from macroscopic parameters, apply at very low values of n. In fact there is no fundamental reason why this should be so.

The value of r_C obtained from the macroscopic density may not be appropriate as the particle size diminishes. In fact, for an individual molecule, the effective radius r_1 can be obtained from gas viscosity measurements. Fortunately these values are usually comparable. For example at 100°C, $r_C = 1.953 \times 10^{-8}$ cm and $r_1 = 2.088 \times 10^{-8}$ cm for H_2O; and at 25°C, $r_C = 2.79 \times 10^{-8}$ cm and $r_1 = 2.76 \times 10^{-8}$ cm for CH_3Cl. Thus for all practical purposes we can assume the average molecular radius independent of particle size. In all our calculations, we use r_C as derived from macroscopic densities, as this is surely correct for large particles.

Table VIII

Comparison of Computed and Experimental Values of $K_{1,1}$ for the Reaction $2C \rightleftarrows C_2$

Molecule M.W.	γ_∞ (dyne/cm)	4×10^{-15} $\pi\gamma_\infty/\kappa T$ (cm^{-2})	$[C]_{vp}$ (atm)	Density (g/cm^3)	$10^8 r_C$ (cm)	$K_{1,1}\{calc\}^a$ (atm^{-1})	$K_{1,1}\{exp\}^b$ (atm^{-1})	$K_{1,1}\{exp\}^c$ (atm^{-1})
			T = 298.1°K					
H$_2$O (18.02)	71.97	21.978	3.1447×10^{-2}	0.99707	1.928	0.0486	0.0349	0.0437
NH$_3$ (17.03)	20.2	6.168	9.7961	0.7710^d	2.061	0.01276^e	0.00237	
CH$_3$Cl (50.49)	15.4	4.7026	5.322	0.9159^f	2.795	0.01018	0.00213	
CH$_3$CHO (44.05)	20.52	6.266	1.2203	0.7834^g	2.814	0.0160	0.0273	
CH$_3$CN (41.05)	28.65	8.7487	0.1168	0.7762	2.757	0.0437	0.2201	
			T = 373.1°K					
H$_2$O (18.02)	58.9	14.371	1.00	0.95838	1.953	0.0129^h	0.00501	0.0144
CH$_3$CHO (44.05)	11.2	2.7326	11.544		~2.90^i	0.01398	0.00591	
CH$_3$CN (41.05)	19.0	4.6357	1.8703	0.6782	2.884	0.0251	0.0371	

^{a} $-\ln\{K_{1,1}[C]_{vp}\} = (4\pi r_C^2 \gamma_\infty / \kappa T 2^{1/3})$.
^{b} $K_{1,1}\{exp\}$ from second virial coefficient data of Rowlinson (1949).
^{c} $K_{1,1}\{exp\}$ from Bolander et al. (1969) using second virial coefficient data of Keyes (1947).
^{d} At 760 Torr pressure.
^{e} $K_{1,1}\{calc\} = 0.00995$ atm^{-1} if the gas viscosity value of 2.181×10^{-8} cm is used for r_C.
^{f} At 293.1°K.
^{g} At 291.1°K.
^{h} $K_{1,1}\{calc\} = 0.00692$ atm^{-1} if the gas viscosity value of 2.088×10^{-8} cm is used for r_C.
^{i} Estimate.

The variation of surface tension with the radius of curvature has been discussed by Tolman (1949). He developed an approximate formula

$$\gamma_n = \gamma_\infty/(1 + 2\delta/r_n) \tag{104}$$

where γ_n is the surface tension of C_n and δ is the distance ($\sim 10^{-8}$ cm) between the surface of tension and the surface at which the superficial density vanishes. As a useful formula, Tolman's formula has three shortcomings:

1. It is applicable only for $r_n \gg 2\delta$.
2. δ is more or less an empirical parameter, since it cannot be measured directly.
3. γ_n does not go to zero in the limit as $n \Rightarrow 1$.

We can, however, utilize Tolman's formula by realizing that it only applies when $r_n \gg 2\delta$. Then expansion by the binomial theorem gives

$$\gamma_n = \gamma_\infty(1 - 2\delta/r_n) \tag{105}$$

Since we require $\gamma_n = 0$ when $n = 1$, we let $2\delta = r_C$ which gives us a workable formula

$$\gamma_n = \gamma_\infty(1 - r_C/r_n) \tag{106}$$

This formula gives the correct dependence for $n = 1$ and for large n. There is no reason to expect it to be accurate for intermediate values of n.

We can test the validity of the approximation by computing $K_{1,1}$ for several substances and comparing with experimental values. When this is done, the fit is not good. A much better fit (by several orders of magnitude in some cases) occurs by using the relationship

$$\gamma_n = \gamma_\infty[1 - (r_C/r_n)^3] = \gamma_\infty[1 - (1/n)] \tag{107}$$

Then
$$\ln([C]_{vp} K_{1,1}) = -4\pi r_C^2 \gamma_\infty/\kappa T(2^{1/3}) \tag{108}$$

With Eq. (108), $K_{1,1}$ has been computed for several molecules for which the experimental values can be deduced from virial coefficient data. Both the computed and experimental values are listed in Table VIII, along with the parameters needed for the computation. For H_2O, virial coefficient data are available from both the work of Rowlinson (1949) and Keyes (1947). The experimental equilibrium constants obtained from the virial coefficient data give somewhat differing results, especially at 373.1°K. The discrepancy between the results is an indication of the experimental uncertainty. The values of $K_{1,1}$ calculated from Eq. (108) agree remarkably well with the values deduced from Keyes data. For the other molecules,

Table IX

Equilibrium Constants $K_{1,1}$ for the Reaction $2C \rightleftarrows C_2$

Compound	Temp. (°K)	γ_∞ (dynes/cm)	ρ (g/cm^3)	$10^8 r_C$ (cm)	$[C]_{vp}$ (atm)	M.W.	$K_{1,1}\{obs\}^a$ (atm^{-1})	$K_{1,1}\{calc\}^b$ (atm^{-1})	$K_{1,1}\{calc\}^c$ (atm^{-1})
H$_2$O	293.1	72.75	0.9992	1.926	0.02307	18.02	8.4×10^{-4}	0.0560	7.23×10^{-5}
CO$_2$	180	21.0	1.01^d	2.585	0.7993	44.01	2.28×10^{-3}	4.47×10^{-3}	1.60×10^{-5}
	200	17.5	1.01^d	2.585	2.414	44.01	1.37×10^{-3}	6.06×10^{-3}	8.87×10^{-5}
	220	14.0	1.01^d	2.585	5.962	44.01	8.36×10^{-4}	7.77×10^{-3}	3.60×10^{-4}
	240	10.5	1.01^d	2.585	12.736	44.01	5.32×10^{-4}	9.50×10^{-3}	1.15×10^{-3}
	260	7.0	1.01^d	2.585	23.98	44.01	3.27×10^{-4}	1.14×10^{-2}	3.10×10^{-3}
	280	3.5	1.01^d	2.585	41.21	44.01	1.90×10^{-4}	1.33×10^{-2}	7.26×10^{-3}
	300	0.2	1.01^d	2.585	66.47	44.01	1.52×10^{-4}	1.46×10^{-2}	1.41×10^{-2}

^{a} $K_{1,1}\{obs\}$ from Leckenby and Robbins (1966).
^{b} $K_{1,1}\{calc\} = [C]_{vp}^{-1} \exp\{-4\pi r_C^2 \gamma_\infty / \kappa T 2^{1/3}\}$.
^{c} $K_{1,1}\{calc\} = [C]_{vp}^{-1} \exp\{-4\pi r_C^2 \gamma_\infty 2^{2/3}/\kappa T\}$.
^{d} At -37°C for liquid CO$_2$.

the agreement is not extremely good, but the calculated and experimental values agree to within a factor of five in all cases. This is not too bad, since equilibrium constants depend on the exponential of the free energy term. Small variations in parameters can have significant effects. For example, if the molecular radius of 2.088×10^{-8} cm for H_2O obtained from gas viscosity measurements at 100°C is used in place of $r_C = 1.953 \times 10^{-8}$ cm, then $K_{1,1}$ is reduced by almost a factor of 2. Likewise for NH_3 at 25°C, use of the gas viscosity value of 2.181×10^{-8} cm for the molecular radius reduces $K_{1,1}$ by 20% to a value somewhat closer to the experimental one.

Experimental observations of dimers were made by Leckenby and Robbins (1966) who adiabatically expanded gases and mass spectrometrically monitored the monomer and dimer concentrations. Their equilibrium data for H_2O and CO_2 are given in Table IX. There are uncertainties in the data, because of the experimental procedure. Probably of most importance is that sampling required passing the gas through an orifice into the low-

Table X

Values of $S_{m,n+m}$ for $4\pi r_C^2 \gamma_\infty / \kappa T = 8.1693$[a]

n	1	2	3	4	6	8	10	12
1	5.970 + 02	1.192 + 02	5.990 + 01	4.031 + 01	2.571 + 01	1.993 + 01	1.688 + 01	1.500 + 01
2	1.154 + 02	1.070 + 01	3.361 + 00	1.636 + 00	6.785 − 01	3.969 − 01	2.742 − 01	2.083 − 01
3	5.611 + 01	3.305 + 00	7.566 − 01	2.907 − 01	8.615 − 02	3.990 − 02	2.314 − 02	1.532 − 02
4	3.655 + 01	1.583 + 00	2.874 − 01	9.200 − 02	2.073 − 02	7.874 − 03	3.918 − 03	2.293 − 0
6	2.194 + 01	6.355 − 01	8.337 − 02	2.039 − 02	3.007 − 03	2.265 − 04	3.179 − 04	1.505 − 04
8	1.612 + 01	3.599 − 01	3.779 − 02	7.621 − 03	8.171 − 04	1.741 − 04	5.428 − 05	2.153 − 0
10	1.302 + 01	2.410 − 01	2.144 − 02	3.731 − 03	3.110 − 04	5.381 − 05	1.407 − 05	4.794 − 0
12	1.110 + 01	1.778 − 01	1.389 − 02	2.148 − 03	1.457 − 04	2.117 − 05	4.760 − 06	1.422 − 0
14	9.780 + 00	1.395 − 01	9.806 − 03	1.376 − 03	7.850 − 05	9.821 − 06	1.938 − 06	5.154 − 0
16	8.820 + 00	1.144 − 01	7.354 − 03	9.503 − 04	4.674 − 05	5.134 − 06	9.025 − 07	2.165 − 0
18	8.087 + 00	9.669 − 02	5.764 − 03	6.938 − 04	2.999 − 05	2.937 − 06	4.657 − 07	1.018 − 0
20	7.508 + 00	8.369 − 02	4.671 − 03	5.284 − 04	2.038 − 05	1.802 − 06	2.605 − 07	5.235 − 0
25	6.475 + 00	6.271 − 02	3.063 − 03	3.052 − 04	9.299 − 06	6.634 − 07	7.883 − 08	1.323 − 0
30	5.788 + 00	5.033 − 02	2.217 − 03	2.001 − 04	5.064 − 06	3.043 − 07	3.084 − 08	4.462 − 0
40	4.919 + 00	3.655 − 02	1.383 − 03	1.078 − 04	2.061 − 06	9.530 − 08	7.548 − 09	8.659 − 1
70	3.744 + 00	2.130 − 02	6.209 − 04	3.750 − 05	4.382 − 07	1.265 − 08	6.379 − 10	4.748 − 1
100	3.228 + 00	1.586 − 02	4.003 − 04	2.097 − 05	1.856 − 07	4.087 − 09	1.585 − 10	9.140 − 1
300	2.254 + 00	7.753 − 03	1.373 − 04	5.062 − 06	2.237 − 08	2.486 − 10	4.916 − 12	1.460 − 1
1×10^3	1.723 + 00	4.536 − 03	6.150 − 05	1.737 − 06	4.514 − 09	2.959 − 11	3.460 − 13	6.092 − 1
3×10^3	1.459 + 00	3.250 − 03	3.731 − 05	8.923 − 07	1.663 − 09	7.822 − 12	6.568 − 14	8.307 − 1
1×10^4	1.288 + 00	2.533 − 03	2.566 − 05	5.419 − 07	7.872 − 10	2.886 − 12	1.890 − 14	1.864 − 1
3×10^4	1.192 + 00	2.169 − 03	2.034 − 05	3.975 − 07	4.946 − 10	1.553 − 12	8.710 − 15	7.358 − 1
1×10^5	1.124 + 00	1.932 − 03	1.710 − 05	3.153 − 07	3.494 − 10	9.772 − 13	4.880 − 15	3.672 − 1
3×10^5	1.085 + 00	1.798 − 03	1.535 − 05	2.730 − 07	2.816 − 10	7.329 − 13	3.406 − 15	2.385 − 1
∞	1.000 + 00	1.451 − 03	1.079 − 05	1.816 − 07	1.164 − 10	3.823 − 13	1.256 − 15	9.335 − 1

[a] Entries give the value and the power of 10; e.g., $1.234 - 06 = 1.234 \times 10^{-6}$. $\gamma_n = \gamma_\infty [1 - (1/n)]$.

RESULTS AT LOW n

pressure mass spectrometer ionization chamber. Thus during sampling, the relative dimer to monomer concentration should decrease and the experimental values for their ratio will be low, this effect being more important at the higher temperatures, where dissociation rates are faster.

For H_2O at 18°C, the experimental value of $K_{1,1}$ is about two orders of magnitude lower than computed from second virial coefficient data or from the theory with $\gamma_2 = \gamma_\infty/2$. If γ_2 is taken as γ_∞, then the theoretical value is reduced by about three orders of magnitude, and if $\gamma_2 = \gamma_\infty[1 - (r_C/r_2)]$, then the theoretical value is very much higher. All in all the value of $\gamma_2 = \gamma_\infty/2$ appears most suitable.

With the CO_2 data, the theoretical computations all give higher values of $K_{1,1}$ than measured since $\gamma_\infty \sim 0$, i.e., we are near the critical temperature of 304.2°K. Presumably the experimental values are too low for the reason discussed previously. As the temperature is reduced the theoretical values fall; those with $\gamma_2 = \gamma_\infty$ falling the most rapidly, and fit the experi-

m									
	14	16	18	20	25	30	40	70	100
373 + 01	1.282 + 01	1.213 + 01	1.160 + 01	1.067 + 01	1.009 + 01	9.396 + 00	8.666 + 00	8.511 + 00	
681 - 01	1.414 - 01	1.225 + 01	1.085 - 01	8.561 - 02	7.194 - 02	5.647 - 02	3.897 - 02	3.265 - 02	
105 - 02	8.453 - 03	6.754 - 03	5.576 - 03	3.817 - 03	2.874 - 03	1.918 - 03	1.003 - 03	7.210 - 04	
493 - 03	1.048 - 03	7.774 - 04	6.013 - 04	3.601 - 04	2.443 - 04	1.397 - 04	5.584 - 05	3.457 - 05	
199 - 05	4.936 - 05	3.202 - 05	2.200 - 05	1.031 - 05	5.763 - 06	2.461 - 06	5.885 - 07	2.727 - 07	
007 - 05	5.305 - 06	3.059 - 06	1.892 - 06	7.112 - 07	3.330 - 07	1.084 - 07	1.592 - 08	5.580 - 09	
964 - 06	9.207 - 07	4.782 - 07	2.691 - 07	8.279 - 08	3.292 - 08	8.326 - 09	7.684 - 10	2.055 - 10	
185 - 07	2.190 - 07	1.035 - 07	5.351 - 08	1.370 - 08	4.686 - 09	9.347 - 10	5.537 - 11	1.139 - 11	
694 - 07	6.515 - 08	2.829 - 08	1.353 - 08	2.927 - 09	8.696 - 10	1.388 - 10	5.379 - 12	8.584 - 13	
481 - 08	2.290 - 08	9.204 - 09	4.098 - 09	7.586 - 10	1.977 - 10	2.556 - 11	6.599 - 13	8.226 - 14	
801 - 08	9.162 - 09	3.429 - 09	1.430 - 09	2.290 - 10	5.278 - 11	5.592 - 12	9.774 - 14	9.584 - 15	
333 - 08	4.063 - 09	1.424 - 09	5.588 - 10	7.818 - 11	1.605 - 11	1.409 - 12	1.692 - 14	1.314 - 15	
852 - 09	7.445 - 10	2.260 - 10	7.750 - 11	8.021 - 12	1.269 - 12	7.225 - 14	3.610 - 16	1.606 - 17	
377 - 10	1.921 - 10	5.162 - 11	1.579 - 11	1.260 - 12	1.586 - 13	6.148 - 15	1.377 - 17	3.632 - 19	
306 - 10	2.434 - 11	5.373 - 12	1.363 - 12	7.082 - 14	6.078 - 15	1.220 - 16	6.600 - 20	6.694 - 22	
727 - 12	5.912 - 13	8.893 - 14	1.559 - 14	3.403 - 16	1.318 - 17	6.439 - 20	1.353 - 24		
098 - 13	6.970 - 14	8.284 - 15	1.155 - 15	1.456 - 17	3.365 - 19	6.374 - 22			
897 - 15	3.041 - 16	1.916 - 17	1.429 - 18	3.918 - 21	2.076 - 23				
463 - 16	4.497 - 18	1.693 - 19	7.564 - 21						
434 - 17	3.170 - 19	8.588 - 21							
509 - 18	4.325 - 20								
484 - 19	1.253 - 20								
771 - 19	4.960 - 21								
279 - 19									
940 - 20									

Table XI

Values of $S_{1,n+1}$ for Various Values of $4\pi r_C^2 \gamma_\infty / \kappa T$ [a]

n	2	4	6	8	10	12	14	16	18	20
1	4.423 + 00	2.163 + 01	1.058 + 02	5.175 + 02	2.531 + 03	1.238 + 04	6.055 + 04	2.962 + 05	1.449 + 06	7.085 + 06
2	2.974	9.736	3.188 + 01	1.044 + 02	3.417 + 02	1.119 + 03	3.663 + 03	1.199 + 04	3.927 + 04	1.286 + 05
3	2.517	6.886	1.884 + 01	5.153 + 01	1.409 + 02	3.856 + 02	1.055 + 03	2.885 + 03	7.892 + 03	2.159 + 04
4	2.285	5.613	1.379 + 01	3.387 + 01	8.319 + 01	2.043 + 02	5.019 + 02	1.233 + 03	3.029 + 03	7.439 + 03
6	2.041	4.407	9.517	2.055 + 01	4.438 + 01	9.584 + 01	2.070 + 02	4.470 + 02	9.652 + 02	2.084 + 03
8	1.907	3.810	7.610	1.520 + 01	3.037 + 01	6.067 + 01	1.212 + 02	2.421 + 02	4.837 + 02	9.663 + 02
10	1.819	3.444	6.518	1.234 + 01	2.335 + 01	4.420 + 01	8.366 + 01	1.583 + 02	2.997 + 02	5.672 + 02
12	1.756	3.193	5.803	1.055 + 01	1.917 + 01	3.485 + 01	6.336 + 01	1.152 + 02	2.093 + 02	3.805 + 02
14	1.708	3.007	5.295	9.322	1.641 + 01	2.890 + 01	5.088 + 01	8.959 + 01	1.577 + 02	2.777 + 02
16	1.669	2.863	4.912	8.426	1.445 + 01	2.479 + 01	4.253 + 01	7.296 + 01	1.252 + 02	2.147 + 02
18	1.637	2.748	4.612	7.740	1.299 + 01	2.180 + 01	3.659 + 01	6.141 + 01	1.031 + 02	1.730 + 02
20	1.610	2.652	4.369	7.197	1.185 + 01	1.953 + 01	3.217 + 01	5.298 + 01	8.728 + 01	1.438 + 02
25	1.558	2.472	3.923	6.226	9.882	1.568 + 01	2.489 + 01	3.950 + 01	6.270 + 01	9.950 + 01
30	1.518	2.343	3.615	5.579	8.608	1.328 + 01	2.050 + 01	3.163 + 01	4.880 + 01	7.530 + 01
40	1.463	2.168	3.211	4.758	7.049	1.044 + 01	1.547 + 01	2.292 + 01	3.400 + 01	5.031 + 01
70	1.374	1.901	2.632	3.643	5.042	6.978	9.658	1.337 + 01	1.851 + 01	2.561 + 01
100	1.327	1.770	2.361	3.150	4.202	5.606	7.479	9.977	1.331 + 01	1.776 + 01
300	1.218	1.487	1.815	2.216	2.705	3.302	4.030	4.920	6.005	7.330
1 × 10³	1.142	1.305	1.491	1.704	1.947	2.225	2.542	2.905	3.319	3.793
3 × 10³	1.097	1.203	1.319	1.447	1.587	1.741	1.910	2.095	2.298	2.520
1 × 10⁴	1.064	1.132	1.204	1.281	1.363	1.450	1.542	1.641	1.745	1.857
3 × 10⁴	1.044	1.090	1.137	1.187	1.239	1.294	1.350	1.410	1.471	1.536
1 × 10⁵	1.029	1.059	1.090	1.122	1.154	1.188	1.223	1.258	1.295	1.333
3 × 10⁵	1.020	1.041	1.062	1.083	1.105	1.127	1.150	1.173	1.196	1.220

[a] Entries give the value and the power of 10; e.g., 1.234 − 06 = 1.234 × 10⁻⁶. $\gamma_n = \gamma_\infty [1 - (1/n)]$.

ments at $\sim 250°K$. With $\gamma_2 = \gamma_\infty/2$, the best fit occurs at $\sim 180°K$. For $\gamma_2 = \gamma_\infty[1 - (r_C/r_2)]$, the theoretical values would be high over the whole experimental region. We see that the best fit is obtained when γ_2 lies between γ_∞ and $\gamma_\infty/2$.

Finally we point out that the calculated values of $K_{1,1}$ for H_2O at 298.1 and 373.1°K, using Eq. (107) to compute γ_2, give an enthalpy of reaction of -3.9 kcal/mole. Quantum-mechanical calculations using the Hartree–Fock approximations give a value between -3.3 and -4.6 kcal/mole at 298.1°K (Kistenmacher et al., 1974).

Equation (108) is not exact because it includes only surface energy terms, and neglects any variation in chemical interaction terms from macroscopic values. It is surprising that it works as well as it does. In fact if the chemical interaction terms are large, as for acetic acid dimers, Eq. (108) fails badly. Thus we conclude, that as a first approximation, Eq. (107) can be used to compute surface tension. It is certainly better than assuming γ_n independent of n or using Tolman's formula or Eq. (106). Of course Eq. (107) is now not going to give Tolman's theoretically derived function for large n. This is, however, not important for us, since at large n, γ_n approaches γ_∞ anyway.

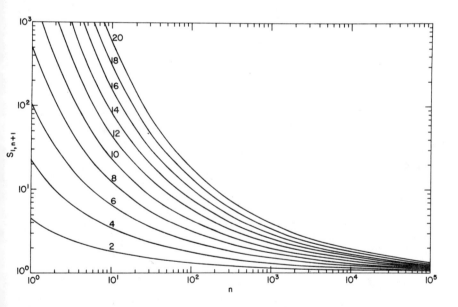

Fig. 11 Log–log plot of $S_{1,n+1}$ versus n for various values of $4\pi r_C^2 \gamma_\infty/\kappa T$ (as indicated on the curves) assuming $\gamma_n = \gamma_\infty[1 - (1/n)]$.

With the use of Eq. (107) to compute γ_n, values of $S_{m,n+m}$ can be computed from the general formula:

$$S_{m,n+m} = \frac{m+n}{[n(n+2m)]^{1/2}} \left[\frac{m^{1/3} + n^{1/3}}{m^{1/3} + (m+n)^{1/3}}\right]^2$$

$$\times \exp\left\{\frac{4\pi r_C^2 \gamma_\infty}{\kappa T}\left[\frac{n+m-1}{(n+m)^{1/3}} - \frac{n-1}{n^{1/3}} - \frac{m-1}{m^{1/3}}\right]\right\} \quad (109)$$

The values have been computed for H_2O at 25.0°C using the parameters in Table VIII. These values of $S_{m,n+m}$ are listed in Table X. Except for $S_{1,2}$ they are always larger than for the approximation that $\gamma_n = \gamma_\infty$ for $n \geq 2$. ($S_{1,2}$ is almost three orders of magnitude smaller.) Also, $S_{2,4}$, $S_{2,5}$, $S_{2,6}$, $S_{3,5}$, and $S_{4,6}$ are >1.0, whereas with the approximation of constant γ, they are all <1.0.

Values of $S_{1,n+1}$ for other values of $4\pi r_C^2 \gamma_\infty/\kappa T$ are given in Table XI. They are all greater than 1.0, and increase with $4\pi r_C^2 \gamma_\infty/\kappa T$. The results also are shown graphically in Fig. 11.

CHAPTER V

HOMOGENEOUS NUCLEATION

We now turn our attention to the problem of a supersaturated vapor. Sooner or later such a vapor must condense. We are interested in the time history of this process. Specifically we ask four questions.

1. What is the time τ_i required for species C_i to begin to appear?
2. At any time, what is the rate of formation of any C_i, and of all C_i with i greater than some minimum value s?
3. How many particles greater than some minimum size C_s are present at any time, i.e., what is $\sum_{n=s}^{\infty} [C_n]$?
4. What is the total mass in particles of size C_s and larger, i.e., what is $\sum_{n=s}^{\infty} n[C_n]$?

Equilibrium Theory

In the equilibrium theory it is assumed that:

Assumption 1 Coagulation is unimportant, and that condensation controls the growth process. Thus the only reactions of concern are

$$C_n + C \rightleftarrows C_{n+1}$$

The justification for this assumption is that just prior to nucleation the concentration of C is so much larger than for any of the other C_n that coagulation rates are too small to have any effect.

Assumption 2 For particles less than or equal to some critical size q the condensation reactions are in equilibrium, but that for $n > q$, $[C_n] = 0$.

Assumption 3 The rate of particle production R_{nucl} is the rate at which C_{q+1} is formed, i.e.,

$$R_{nucl} = k_{1,q}[C][C_q] \qquad (110)$$

Since $[C_q]$ is at its equilibrium concentration, an expression for $[C_q]$ in terms of $[C]$ can be developed. Consider the sequence of reactions, all at equilibrium

$$2C \rightleftarrows C_2$$
$$C_2 + C \rightleftarrows C_3$$
$$C_3 + C \rightleftarrows C_4$$
$$\vdots$$
$$C_{q-1} + C \rightleftarrows C_q$$

The sum of these reactions is

$$qC \rightleftarrows C_q$$

and the equilibrium constant K_{nucl} is

$$K_{nucl} = \prod_{j=1}^{q-1} K_{1,j} \qquad (111)$$

The expression for $K_{1,j}$ is given by generalizing Eq. (78) for γ_n not constant

$$K_{1,j} = [C]_{vp}^{-1} \exp\left\{\frac{-4\pi r_C^2}{\kappa T}[\gamma_{j+1}(j+1)^{2/3} - \gamma_j j^{2/3}]\right\} \qquad (112)$$

Thus

$$K_{nucl} = \frac{1}{[C]_{vp}^{q-1}} \exp\left\{\frac{-4\pi r_C^2 \gamma_q q^{2/3}}{\kappa T}\right\} \qquad (113)$$

Also

$$[C_q] = K_{nucl}[C]^q \qquad (114)$$

so that

$$R_{nucl} = k_{1,q} K_{nucl}[C]^{q+1} \qquad (115)$$

EQUILIBRIUM THEORY 59

Again, since all the reactions are at equilibrium, $k_{1,q} = Z_{1,q}$ and

$$R_{nucl} = Z_{1,q} \frac{[C]^{q+1}}{[C]_{vp}^{q-1}} \exp\left\{\frac{-4\pi r_C^2 \gamma_q q^{2/3}}{\kappa T}\right\} \tag{116}$$

or, since $S_1 \equiv [C]/[C]_{vp}$

$$R_{nucl} = Z_{1,q} S_1^{q+1} [C]_{vp}^2 \exp\left\{\frac{-4\pi r_C^2 \gamma_q q^{2/3}}{\kappa T}\right\} \tag{117}$$

Lothe and Pound (1962) (also see Hirth and Pound, 1963, p. 20) have argued that additional entropy and energy terms are needed because the small particles produced have additional entropies and energies over those of infinitely large particles. The entropy terms arise because the free volumes for translation and rotation of n molecules in the bulk liquid are not as great as the corresponding free volumes for a cluster of n molecules in the vapor, and that these differences in free volumes are probably not compensated by the macroscopic surface entropy of the bulk liquid. These views were supported by Courtney (1968). Reiss and his co-workers (Reiss and Katz, 1967; Reiss et al., 1968) disagreed that rotation played any role, but felt that translational terms were important, because the center of mass could move around within the boundary of a cluster of molecules. We have circumvented the energy problem in the development used here by placing all extra energy terms in the surface energy γ_n and letting γ_n be a function of n. How important are the entropy terms? From the experimental equilibrium constants in Table VIII, the entropy for the reaction

$$2C \rightleftarrows C_2$$

can be computed to be -18, -22, and -21, cal/mole-°K, respectively, for H_2O, CH_3CHO, and CH_3CN. These values are sufficiently close to the Trouton's rule value of 21 cal/mole-°K, so that the difference is unimportant. Thus we will neglect the Lothe–Pound entropy terms and see that the experimental data can be reasonably well fitted without them.

Constant Critical Size q

Let us assume that q is a constant at any temperature, independent of the supersaturation. Then a plot of log R_{nucl} vs ln S_1 should give a straight line with slope $q + 1$ and intercept $Z_{1,q}[C]_{vp}^2 \exp\{-4\pi r_C^2 \gamma_q q^{2/3}/\kappa T\}$.

Data for the rate of nucleation of H_2O vapor at $-5°C$ have been obtained by Allard and Kassner (1965) as a function of supersaturation. The log–log plot is shown in Fig. 12. All the data can be fitted by a straight line whose slope gives $q + 1 = 38$. The intercept is 6×10^{-26} particles/cm³-sec which is very much smaller than the value of 1.12×10^{-20} particles/cm³-sec calculated from $Z_{1,q}[C]^2 \exp\{-4\pi r_C^2 \gamma_q q^{2/3}/\kappa T\}$.

The simple rate law, Eq. (117), works for more complex systems than H_2O vaporization. The results for several systems were reviewed by Heicklen and Luria (1975). The summary of the results is given in Table XII. It is interesting to note that except for H_2O, the critical size particle needed for nucleation contains less than eight basic units for the other five systems for which we have data.

Variable Critical Size q

Equation (116) overestimates the rate because, when $[C_{q+1}] = 0$, $[C_q]$ has not yet reached its equilibrium value, regardless of what the value is for q. In the example for H_2O discussed previously the calculated value for the nucleation rate coefficient was a factor of 1.9×10^5 larger than the experimental value. Equation (117) can, however, be used to compute an

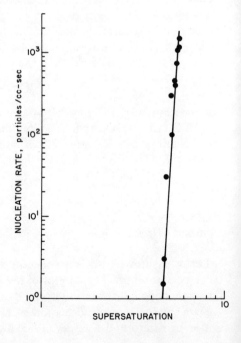

Fig. 12 Log–log plot of the particle number growth rate for H_2O droplets versus the supersaturation of H_2O vapor at $-5°C$. Data of Allard and Kassner (1965). From Heicklen and Luria (1975) with permission of John Wiley and Sons.

Table XII
Experimental Values of q for Various Systems[a]

Nucleation reaction	Temp. (°C)	q	Reference[b]
$q(C_3H_4S_2O_3)_3 \to (C_3H_4S_2O_3)_{3q}$	29	2.6	Luria et al. (1974a)
qPb $\to$ Pb$_q$	517–723	2.0	Homer and Prothero (1973)
qNH$_3$ + qHNO$_3$ $\to$ (NH$_4$NO$_3$)$_q$	25	7.0	Olszyna et al. (1974)
$qC_5H_8SO \to (C_5H_8SO)_q$	29	5.7	Luria et al. (1974b)
qNH$_3$ + qHCl $\to$ (NH$_4$Cl)$_q$	25	7.5	Twomey (1959); Countess and Heicklen (1973)
qH$_2$O $\to$ (H$_2$O)$_q$	−5	37	Allard and Kassner (1965)

[a] From Heicklen and Luria (1975).
[b] Reference for experimental data.

upper bound for the nucleation rate. Obviously the value of q which gives the lowest upper bound will be the most useful. This value is found by taking dR_{nucl}/dq and setting it equal to zero. If we ignore the small dependence of $Z_{1,q}$ and γ_q on q, this value of q is

$$q = (8\pi r_C^2 \gamma_q / 3\kappa T \ln S_1)^3 \tag{118}$$

It should be observed that the value of q computed from Eq. (118) is a function of the supersaturation (i.e., it is not a constant at any temperature) and corresponds to the particle size at steady-state supersaturation, i.e., C_q is evaporating at the same rate at which it is growing by condensation of vapor on it.

When Eq. (118) is substituted into Eq. (117) the rate equation for nucleation becomes

$$\ln R_{nucl} = \ln(Z_{1,q}[C]_{vp}^2 S_1) - \frac{1}{2(\ln S_1)^2}\left(\frac{8\pi r_C^2 \gamma_q}{3\kappa T}\right)^3 \tag{119}$$

which was first proposed by Volmer and Weber (1926).

Figure 13 is a replot of the nucleation rate data of Allard and Kassner (1965) for H$_2$O at −5°C. Also shown on the same plot is the nucleation rate computed from the exact Volmer–Weber theory, i.e., R$_{nucl}$ was calculated from Eq. (117) for all values of q and the minimum value used. The values of q obtained in this way ranged from 98 to 45 as S_1 increased from 4 to 6. These values are larger than the value of 37 obtained from the "constant q" theory. The computed values for R$_{nucl}$ are still very much higher than the observed values, the discrepancy being about two orders of magnitude at low supersaturations and four to five orders of magnitude

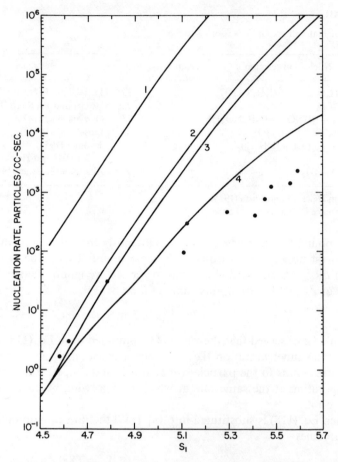

Fig. 13 Nucleation rate of H_2O droplets at 268.1°K versus the water supersaturation ratio S_1 from the data of Allard and Kassner (1965) as reported by Kassner and Schmitt (1966). The four curves represent computations from 1. equilibrium theory (Volmer–Weber), 2. steady-state theory (Becker–Döring) 3. modified steady-state theory, and 4. kinetic theory (Frenkel–Courtney). The values for the parameters used in the calculations are given in Table XIII.

at high supersaturations. This is, however, an improvement over the discrepancy of a factor of 1.9×10^5 computed from the theory with constant q.

Induction Time

If we assume that there is some time τ_i when the equilibrium condition is valid for all C_n smaller than some size C_i, but that $[C_i] = 0$, then τ_i can be evaluated. This would be the induction time to produce C_i. This time is given by

$$\tau_i = \sum_{n=q}^{i-1} [C_n]/R_{nucl} \qquad (120)$$

where it has been assumed that the time required to achieve equilibrium for particles smaller than or equal to C_q is negligible. (The rates of formation of these particles are greater than R_{nucl}). Equation (120) is the integrated form of the more general equation

$$\int_0^{\tau_i} R_{nucl}\, dt = \sum_{n=q}^{i-1} [C_n] \qquad (121)$$

In order to obtain Eq. (120), it was assumed that R_{nucl} was independent of time, which means that $[C]$ must be a constant. If $[C]$ is not constant as the particles are produced, the situation is considerably more complex, and an equilibrium theory has no significance.

For the case of constant $[C]$, R_{nucl} is constant, and the number and mass of particles increases continually. Since the equilibrium concentration of $[C_n]$ is given by

$$[C_n] = S_1^n [C]_{vp} \exp\left\{\frac{-4\pi r_C^2 \gamma_n}{\kappa T} n^{2/3}\right\} \qquad (122)$$

then

$$\tau_i = \frac{[C]_{vp}}{R_{nucl}} \sum_{n=q}^{i-1} S_1^n \exp\left\{\frac{-4\pi r_C^2 \gamma_\infty (n-1)}{\kappa T}\, \frac{1}{n^{1/3}}\right\} \qquad (123)$$

Steady-State Theory

Becker and Döring (1935) modified the equilibrium theory by replacing the equilibrium assumption 2 with a steady-state assumption. Thus C_q is not in equilibrium with C, but all the C_n for $n \leq q$ are in a steady state. The concentration of C_{q+1} is still taken as zero.

The series of equations to be solved is

$$0 = k_{1,n-1}[C][C_{n-1}] + k_{-1,n}[C_{n+1}] - k_{-1,n-1}[C_n] - k_{1,n}[C][C_n] \qquad (124)$$

for $2 \leq n \leq q$, remembering that $[C_{q+1}] = 0$. When this is done, the expression for $[C_q]$ becomes

$$[C_q] = \frac{[C]^q \prod_{j=1}^{q-1} K_{1,j}}{1 + k_{1,q} \sum_{j=1}^{q-1} \left(\frac{[C]^j \prod_{i=1}^{j-1} K_{1,q-i}}{k_{-1,q-j}} \right)} \tag{125}$$

The product terms are

$$\prod_{i=1}^{q-1} K_{1,j} = \frac{1}{[C]_{vp}^{q-1}} \exp\left\{ \frac{-4\pi r_C^2 \gamma_q q^{2/3}}{\kappa T} \right\} \tag{126}$$

$$\prod_{i=1}^{j-1} K_{1,q-i} = \frac{1}{[C]_{vp}^{j-1}} \exp\left\{ \frac{-4\pi r_C^2}{\kappa T} [\gamma_q q^{2/3} - \gamma_{q-j}(q-j)^{2/3}] \right\} \tag{127}$$

so that the nucleation rate becomes

$$R_{nucl} = \frac{(S_1)^{q+1}[C]_{vp}^2 \exp\left\{ \frac{-4\pi r_C^2 \gamma_q}{\kappa T} q^{2/3} \right\}}{\frac{1}{k_{1,q}} + \sum_{j=1}^{q-1} \left(\frac{(S_1)^j [C]_{vp}}{k_{-1,q-j}} \exp\left\{ \frac{-4\pi r_C^2}{\kappa T} [\gamma_q q^{2/3} - \gamma_{q-j}(q-j)^{2/3}] \right\} \right)} \tag{128}$$

Replacing $k_{-1,q-j}$ by its equivalent form

$$k_{-1,q-j} = k_{1,q-j}[C]_{vp} \exp\left\{ \frac{-4\pi r_C^2}{\kappa T} [\gamma_{q-j+1}(q-j+1)^{2/3} - \gamma_{q-j}(q-j)^{2/3}] \right\} \tag{129}$$

gives

$$R_{nucl} = \frac{k_{1,q}(S_1)^{q+1}[C]_{vp}^2 \exp\left\{ \frac{-4\pi r_C^2 \gamma_q}{\kappa T} q^{2/3} \right\}}{1 + k_{1,q} \sum_{j=1}^{q-1} \frac{(S_1)^j}{k_{1,q-j}} \exp\left\{ \frac{-4\pi r_C^2}{\kappa T} [\gamma_q q^{2/3} - \gamma_{q-j+1}(q-j+1)^{2/3}] \right\}} \tag{130}$$

The numerator of the right-hand side of the equation is just the equilibrium theory nucleation rate. The denominator is a correction term which reduces this rate. Typically the reduction is about two orders of magnitude. This rate must still be an overestimate, however, as will be shown in the next section.

Equation (130) has been used to compute the steady-state nucleation rate for H_2O at $-5°C$ using the value of q at each supersaturation S_1 which corresponds to the steady-state supersaturation $S_{1,q}$, i.e., the value of q is such that the deposition rate of C on C_q equals the evaporation rate of C from C_q. This is the usage of the equation as formulated by Becker and Döring and the one in general use today. The results are shown in Fig. 13 as curve 2. It can be seen that this curve reduces the equilibrium theory rate by two orders of magnitude. Curve 2 lies only slightly above the data points at low supersaturations, but still greatly overestimates the nucleation rate at high supersaturations.

Modified Steady-State Theory

The general equations to be considered are

$$\frac{d[C_n]}{dt} = k_{1,n-1}[C][C_{n-1}] + k_{-1,n}[C_{n+1}] - (k_{1,n}[C] + k_{-1,n-1})[C_n] \quad (131)$$

Let us consider the situation in which particles are considered to be species larger than C_s where s is significantly greater than q. Just at the time of onset of nucleation, $[C_n] = 0$ for all $n > s$. For constant $[C]$, the concentrations of all the C_n ($1 < n \leq s$) start at zero, or some small value, and increase with time to their steady-state values. Thus if we assume the steady-state condition for all C_n with $1 < n \leq s$, we overestimate the concentration of C_s, and thus we overestimate $d[C_{s+1}]/dt$, i.e., the initial nucleation rate.

The steady-state concentration of C_s for $[C_{s+1}] = 0$, is

$$[C_s] = \frac{(S_1)^s[C]_{vp}\exp\left\{\dfrac{-4\pi r_C^2 \gamma_s s^{2/3}}{\kappa T}\right\}}{1 + k_{1,s}\sum_{j=1}^{s-1}\dfrac{(S_1)^j}{k_{1,s-j}}\exp\left\{\dfrac{-4\pi r_C^2}{\kappa T}[\gamma_s s^{2/3} - \gamma_{s-j+1}(s-j+1)^{2/3}]\right\}}$$

(132)

and the rate of nucleation is

$$R_{nucl} = d[C_{s+1}]/dt = k_{1,s}S_1[C]_{vp}[C_s] \quad (133)$$

The value of s is arbitrary, since the minimum particle size is a matter of definition. R_{nucl} is, however, a well-defined parameter independent of time for constant $[C]$ (assuming reactions of all C_n with C_m are negligible for $n,m > 1$). Since the steady-state concentration of C_s will always overestimate R_{nucl}, we should choose s to minimize the calculated value of R_{nucl}.

We anticipate, in advance, that s will be a large number, so that the 1 in the denominator of expression (132) for $[C_s]$ is negligible. Then the initial nucleation rate Eq. (133) becomes

$$(d[C_{s+1}]/dt)^{-1} = \sum_{m=0}^{s} \frac{\exp\{4\pi r_C^2 \gamma_m m^{2/3}/\kappa T\}}{S_1{}^m [C]_{vp}^2 k_{1,m-1}} \tag{134}$$

The right-hand side of Eq. (134) is the sum of positive terms. Thus the larger the value of s, the larger is the right-hand side of the equation and the smaller the computed value for the nucleation rate. Thus the optimum (lowest) value for the computed nucleation rate occurs with $s = \infty$. The function that is being summed in Eq. (134) goes through a sharp maximum for $m = q$, and is approximately symmetric for values of m near q. Thus for $s \gg q$, the sum is almost exactly twice (rate is $\frac{1}{2}$) that for $s = q$.

As an example of the application of the steady-state theory, Fig. 14 shows the initial rate of production of C_{s+1} as a function of s for H_2O vapor at 268.1°K with $S_1 = 6.0$. For comparison purposes the rate assuming equilibrium concentrations for C_n ($n \le s$) is also given. The equilibrium theory computation shows that the minimum rate occurs for $s = q$ which is 47 in this example. Since the rate is always overestimated by this theory, this minimum value is the best value (i.e., the least upper bound).

The steady-state theory gives a much slower rate than equilibrium theory, but about twice as high for $s = q$ as for $s > 70$. Since the rate does not change much with s for $s > 70$, it is not necessary to perform an infinite summation. Generally if the summation is performed for $s = 2q$, the lower limiting rate is obtained, and this value, referred to as the modified steady-state rate, is almost exactly $\frac{1}{2}$ the value obtained by the classical steady-state theory of Becker and Döring using $s = q$.

Curve 3 of Fig. 13 gives the modified steady-state theory nucleation rate for H_2O at 268.1°K and various supersaturations. It lies about a factor of two lower than the classical steady-state values. It exactly fits the data points at low supersaturations, but still gives high values at large supersaturations.

Induction Time

The steady-state theory is not convenient for computing induction times, because of the general complexity in evaluating each of the $[C_n]$ to be used in Eq. (120). A lower limiting value, however, applicable at high supersaturations, can be obtained by assuming that all the $k_{-1,n}$ are negligible. In that case, the steady-state values for $[C_n]$ ($n \le s$) are

$$[C_n] = \frac{(S_1)^n [C]_{vp} \exp\left\{\dfrac{-4\pi r_C^2 \gamma_n n^{2/3}}{\kappa T}\right\}}{1 + k_{1,n} \sum_{j=1}^{n-1} \dfrac{(S_1)^j}{k_{1,n-j}} \exp\left\{\dfrac{-4\pi r_C^2}{\kappa T}[\gamma_n n^{2/3} - \gamma_{n-j+1}(n-j+1)^{2/3}]\right\}} \tag{135}$$

Equation (135) is analogous to Eq. (132), but it applies for all $n \leq s$, because of the assumption that all the $k_{-1,n}$ are negligible. With this expression for $[C_n]$ we have seen that

$$R_{\text{nucl}} \leq S_1[C]_{\text{vp}} k_{1,n}[C_n] \qquad (136)$$

Thus

$$\tau_s = \sum_{n=1}^{s-1} \frac{[C_n]}{R_{\text{nucl}}} \geq \frac{1}{S_1[C]_{\text{vp}}} \sum_{n=1}^{s-1} \frac{1}{k_{1,n}} \qquad (137)$$

If we further assume that

$$k_{1,n} \approx Z_{1,n} \approx \left(\frac{8\pi\kappa T}{m_C}\right)^{1/2} r_C^2 n^{2/3} \qquad (138)$$

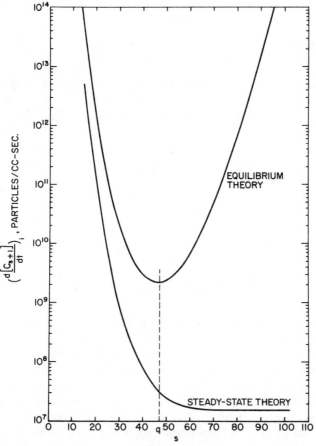

Fig. 14 Semilog plot of $d[C_{s+1}]/dt$ versus s for H_2O at a supersaturation S_1, of 6.0 at 268.1°K. The calculations are done for equilibrium and steady-state theory assuming all particles smaller than C_{s+1} are in equilibrium or steady state, respectively, and $[C_{s+1}] = 0$.

and replace the summation by an integration, then

$$\tau_s \gtrsim 3(s-1)^{1/3}/S_1[C]_{vp}(8\pi\kappa T/m_C)^{1/2} r_C^2 \qquad (139)$$

Allen (1968) found in the nucleation of water vapor in He by adiabatic expansion to $-5°C$ that the vapor pressure dropped 10% from its critical supersaturation value at $-5°C$ in 0.05 to 0.12 sec for all conditions and that about 10 drops of H_2O/cm^3 were produced. Since the vapor pressure at $-5°C$ is 3.17 Torr (1.142×10^{17} molecules/cm^3), $s/S_1 \simeq 1\%$ of that number for the time interval. Also $r_C = 1.926 \times 10^{-8}$ cm, and the values of S_1 ranged from ~ 4.5 to ~ 6.5. Taking $S_1 = 5$ permits us to compute $\tau_s = 0.014$ sec from Eq. (139). As expected the computed value lies below the observed values, but only by a factor of 4 to 8.

Kinetic Theory

In order to solve the problem properly it is necessary to consider the time history of the system. For mathematical simplicity we shall, as in the preceding discussion, ignore the coagulation terms. Thus the generalized rate laws become

$$\frac{d[C]}{dt} = Q + k_{-1} - k_1[C] + 2k_{-1,1}[C_2] - 2k_{1,1}[C]^2$$

$$+ \sum_{n=2}^{\infty} (k_{-1,n}[C_{n+1}] - k_{1,n}[C][C_n]) \qquad (140)$$

$$\frac{d[C_n]}{dt} = k_{-n} - k_n[C_n] + k_{1,n-1}[C][C_{n-1}] + k_{-1,n}[C_{n+1}]$$

$$- k_{1,n}[C][C_n] - k_{-1,n-1}[C_n], \quad n > 1 \qquad (141)$$

In the most general case, Q and all the rate coefficients can be time-dependent functions. In particular the rate coefficients all depend on temperature, so that if the temperature is reduced in order to initiate the nucleation, then all the rate coefficients are time dependent. In addition the coefficients k_{-n} represent physical sources such as pressure changes, wall disintegrations, or flow rate changes, and they can be time dependent even at constant temperature. A general set of solutions has not been obtained for the n-coupled equations. At Penn State University, R. G. de Pena is now programming them for constant-temperature conditions. It is also hoped to include the coagulation terms, if computer time permits.

In spite of the fact that a general solution has not been effected yet, several investigators have examined these equations under particular conditions. Many investigators (Kantrowitz, 1951; Probstein, 1951; Collins, 1955; and Wakeshima, 1954) followed the idea of Frenkel (1955), who converted the right-hand side of Eq. (141) to a continuous function in $[C_n]$. Then they made various approximations to solve the equations. Goodrich (1964b) went to considerable length to make the mathematical equations more tractable. He was able, however, to find an analytical solution only for the artificial case of constant $[C]$, constant temperature, all the $k_{1,n}$ equal and all the $k_{-1,n}$ equal.

Courtney (1962a) worked out the exact solutions for $n \leq 110$ on a high-speed computer based on the following assumptions:

(1) $[C_n]$ is its equilibrium concentration for $n \leq 19$;
(2) rate coefficients for condensation are the collision coefficients $Z_{1,n}$;
(3) $[C]$ = constant; and
(4) constant temperature.

A typical set of results of Courtney's computer computations are shown in Fig. 15. There is an induction period before any C_n appears. Then $[C_n]$ reaches its steady-state concentration in a step function and finally increases more slowly to its equilibrium value as higher C_n are produced.

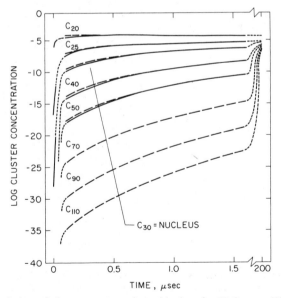

Fig. 15 Variation of cluster concentration with time for H_2O at $-60°C$ and $S_1 = 20$. From Courtney (1962a) with permission of the American Institute of Physics.

Courtney (1962b) extended his work to the situation of constant total mass, i.e., the concentration of C depletes as the higher C_n are formed. Typical result of the supersaturation (S_1)-time curves are shown in Fig. 16. The higher the supersaturation, the faster the nucleation rate, and thus the faster the depletion.

The computed rates of nucleation for H_2O at $-5°C$ from the kinetic theory as developed by Courtney are compared to the data of Allard and Kassner (1965) in Fig. 13. The kinetic theory computation (curve 4) gives the best fit to the data, but still the computed values are about one order of magnitude too high at the largest supersaturations. The reason for the discrepancy may be twofold:

(1) At the higher supersaturations, the vapor is depleted much faster (see Fig. 16) which reduces the experimental rate. The calculation assumed constant [C].

(2) At higher supersaturations, the concentrations of C_n become relatively larger, and this effect is more pronounced for higher n. Coagulation terms may be important and thus reduce the measured rates.

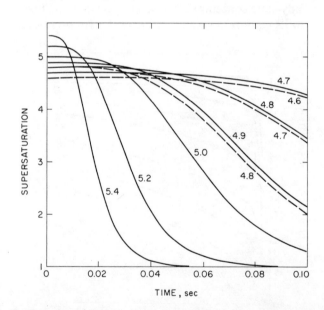

Fig. 16 Condensation of liquid water versus time for various initial supersaturations at $-5°C$. The solid line represents the correct Frenkel nucleation; the dashed line, Frenkel nucleation. From Courtney (1962b) with permission of the American Institute of Physics.

Effect of Diluent

Allen and Kassner (1969) adiabatically expanded He or Ar saturated with H_2O vapor initially at 22.5°C to reduce the temperature and produce various supersaturations of H_2O vapor. Thus nucleation was induced, and the rates measured. Their results are shown graphically in Fig. 17. There are two interesting effects to be noted here:

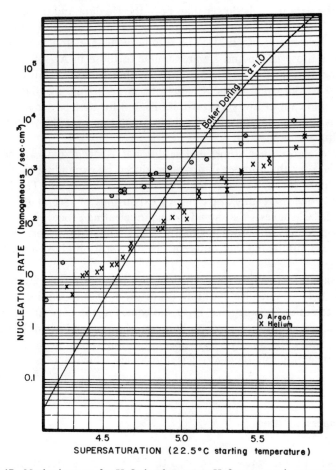

Fig. 17 Nucleation rate for H_2O droplets versus H_2O vapor peak supersaturation from the adiabatic expansion of argon or helium saturated with H_2O vapor and initially at 22.5°C. The curve is the theoretical calculation based on steady-state (Becker–Döring) theory for accommodation coefficients α of unity. From Allen and Kassner (1969) with permission of Academic Press.

(1) The data for Ar lie considerably above those for He, even though theory indicates that the diluent should have no effect.

(2) At low supersaturations the data lie above the theoretical curve of the Becker–Döring steady-state theory even though that theory should predict an upper limit to the nucleation rate.

The data of Allard and Kassner (1965) with air as a diluent are consistent with the He data for supersaturations above 4.8. At lower supersaturations, however, the data in air lie lower than both the He data and the theoretical curve. The data have been extended to other inert gases by Biermann (1971) and Kassner et al. (1971). The results in Ne are similar to those in He, whereas those in Kr and Xe are similar to those in Ar. The enhancement in the rates in the inert gases suggests that some nucleating process in addition to the usual homogeneous one is occurring. For He and Ne it is only significant at low supersaturations, but for the heavier inert gases, the effect is more pronounced and enhances the rate over the whole range of supersaturations.

Further evidence that something additional is occurring can be seen from the data for the number of drops produced by expanding to different supersaturation levels. For He this data is shown in Fig. 18. The knee in the curves, which is readily apparent for expansions of saturated vapor at 22.5 and 12.5°C, suggests that an additional nucleating process is important at the lower supersaturations. In more recent studies (Kassner, 1975) the knee has been eliminated, but different rates are still obtained for different inert gases.

An explanation for this phenomenon has been proposed by Kassner et al. (1971). They visualize clathrate cages of H_2O surrounding the inert gas. An interaction between the H_2O molecules in the cage and the inert gas molecule is established through hydrogen bonding which stabilizes the cage, i.e., lowers the steady-state supersaturation. Thus we have heterogeneous nucleation on the inert gas molecules which act as nucleating centers.

Critical Supersaturation

A number of investigators have used the Wilson cloud chamber to measure the onset of nucleation. In these experiments, a vapor initially at its vapor pressure at room temperature is cooled by adiabatic expansion until condensation is observed. From the expansion ratio, the final temperature and supersaturation can be obtained. The results of such investigations are summarized in Table XIII. Additional data for H_2O have been obtained

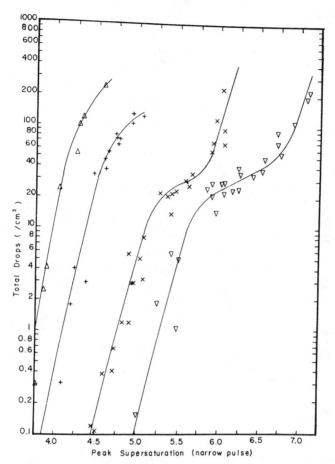

Fig. 18 Nucleation of water droplets versus H_2O vapor peak supersaturation from the adiabatic expansion of a helium atmosphere saturated with water vapor at four different initial temperatures: ▽, 12.5°C; ×, 22.5°C; +, 32.5°C; and △, 41.0°C. From Allen and Kassner (1969) with permission of Academic Press.

by Sander and Damköhler (1943) and Madonna et al. (1961) who have extended the measurements to 200°K. Unfortunately the two sets of data conflict. Sander and Damköhler found an increase in the log of the critical supersaturation with reciprocal temperature. Madonna et al. found no increase in critical supersaturation as the temperature was reduced to about 200°K, and then a fall-off in the critical supersaturation with further reduction in temperature. It is also difficult to estimate γ_∞ and r_C for supercooled H_2O at these low temperatures, so that theoretical computations are difficult. Therefore these data have been omitted from the table.

In Table XIII are also listed the nucleation rates as computed from equilibrium theory, steady-state theory, and modified steady-state theory. It is somewhat difficult to compare the theoretical and experimental data without knowing what nucleation rate corresponds to the onset of nucleation. Generally this value is considered to be about 1 particle/cm³-sec

Table XIII

Supersaturation at Which Homogeneous Condensation Occurs for Various Liquids

Compound	M. W.	Temp. (°K)	γ_∞ (dynes/cm)	$[c]_{vp}$ (atm)	ρ (g/cm³)	$10^8 r_c$ (cm)
H_2O	18.02	263.7	77.2	2.9658×10^{-3}	0.999	1.92
CH_3OH	32.04	270.0	24.9	3.1796×10^{-2}	0.815	2.49
C_2H_5OH	46.07	273.2	23.8	1.5550×10^{-2}	0.81	2.82
$n-C_3H_7OH$	60.11	270.4	25.3	3.6306×10^{-3}	0.821	3.07
$i-C_3H_7OH$	60.11	264.7	23.5	5.6394×10^{-3}	0.80	3.09
$n-C_4H_9OH$	74.12	270.2	26.4	1.1440×10^{-3}	0.83	3.28
CH_3NO_2		252.2				
H_2O	18.02	255	78.8	1.4664×10^{-3}	0.997	1.92
		268.1	76.4	4.1618×10^{-3}	0.999	1.92
Ethyl Acetate	88.12	244.5	32.1	4.4131×10^{-3}	0.96	3.31
Methyl Butyrate	102.13	255.9	28.4	2.8007×10^{-3}	0.94	3.50
Methyl Isobutyrate	102.13	254.2	27.2	4.3354×10^{-3}	0.94	3.50
Propyl Acetate	102.13	258.2	27.5	3.3122×10^{-3}	0.93	3.51
Ethyl Propionate	102.13	250.5	28.3	2.0371×10^{-3}	0.95	3.49
Formic Acid	46.03	225.5	44.3	3.9361×10^{-5}	1.30	2.41
Acetic Acid	60.05	245.8	26.5	5.9576×10^{-4}	1.10	2.78
Propionic Acid	74.08	255.5	29.7	2.2124×10^{-4}	1.03	3.05
Butyric Acid	88.12	248.8	29.9	1.8147×10^{-5}	1.005	3.26
Isobutyric Acid	88.11	253.0	27.6	7.8899×10^{-5}	0.99	3.27
Isovaleric Acid		264.6				
H_2O	18.02	320.1	68.40	1.0474×10^{-1}	0.98940	1.93
		292.2	72.94	2.1680×10^{-2}	0.998385	1.92
		276.3	75.1	7.5868×10^{-3}	0.999968	1.92
		256.7	78.7	1.6891×10^{-3}	0.997	1.92
		246.7	80.5	7.1500×10^{-4}	0.996	1.92

[a] *References:* 1. Volmer and Flood (1934); 2. Wilson (1897); 3. Allard and Kassner (1965); 4. Laby (5. Powell (1928).
[b] From equilibrium theory.
[c] From steady-state theory.
[d] From modified steady-state theory.

because of experimental reasons. The minimum detectable number by most methods is $\sim 10^3$ particles/cm^3 (clean air has a background count of ~ 300 particles/cm^3). The lifetime to diffusional loss in most instruments is of the order of 10^3 sec. Thus a rate of 1 particle/cm^3-sec would correspond to minimum detection. Presumably this value applies to the work of Laby

S_1{obs}	Reference[a]	Critical size (q)	R_{nucl}[b] (cm^{-3}/sec)	R_{nucl}[c] (cm^{-3}/sec)	R_{nucl}[d] (cm^{-3}/sec)
4.85 ± 0.08	1	73	68.78	0.9845	0.482
3.20 ± 0.1	1	28	2.76 x 10^{21}	7.85 x 10^{19}	4.074 x 10^{19}
2.34 ± 0.05	1	122	1.338 x 10^5	2.225 x 10^3	1.10 x 10^3
3.05 ± 0.05	1	111	0.606	0.009376	0.0046
2.80 ± 0.07	1	127	0.04317	6.477 x 10^{-4}	∿3 x 10^{-4}
4.60 ± 0.13	1	74	51.14	0.7376	0.36
6.05 ± 0.15	1				
8.0	2	38	4.04 x 10^9	5.655 x 10^7	2.71 x 10^7
4.55	3	76	104.8	1.5485	0.75
10	4	56	4.23	3.814 x 10^{-2}	0.0209
6	4				
6	4				
5.7	4				
8.6	4	59	0.606	6.43 x 10^{-3}	3.0 x 10^{-3}
37	4	8	1.846 x 10^{19}	1.543 x 10^{17}	7.83 x 10^{16}
13.5	4	8	8.48 x 10^{21}	1.773 x 10^{20}	8.168 x 10^{19}
10	4	24	3.193 x 10^{13}	4.667 x 10^{11}	2.27 x 10^{11}
25	4	15	1.300 x 10^{14}	1.100 x 10^{12}	5.59 x 10^{11}
19	4	15	7.33 x 10^{15}	7.93 x 10^{13}	3.96 x 10^{13}
9	4				
2.87	5	97	1.182 x 10^7	2.00 x 10^5	9.89 x 10^4
3.74	5	78	9.819 x 10^5	1.572 x 10^4	7.6 x 10^3
5.07	5	54	4.971 x 10^8	7.918 x 10^6	3.9 x 10^6
7.80	5	39	4.864 x 10^9	6.525 x 10^7	3.30 x 10^7
8.95	5	39	1.184 x 10^8	1.403 x 10^6	7.25 x 10^5

(1908), Volmer and Flood (1934), and Allard and Kassner (1965). The onset of nucleation reported by Wilson (1897) and by Powell (1928) was the "cloud limit." Presumably this would involve considerably more particles and would correspond to $R_{nucl} \sim 10^3$ particle/cm^3-sec.

The H$_2$O data of Volmer and Flood and Allard and Kassner are consistent with the theory. For the data of Wilson and of Powell the computed values are generally higher than expected. The two esters for which computations were made show R_{nucl} consistent with experiment. The theory gives very poor results for CH$_3$OH, somewhat better results for C$_2$H$_5$OH, and satisfactory results for the higher alcohols. With the organic acids, however, there is a complete breakdown between theory and experiment. The organic acids are always anomolous because of the high association in the vapor. (The dimer of CH$_3$COOH is often more prevalent than the monomer.) This high degree of association is the result of chemical bonding which is not taken into account in the theory.

The most recent data on H$_2$O nucleation in H$_2$ or He has been obtained by Heist and Reiss (1973), who used a thermal diffusion chamber. They obtained critical supersaturations between 285 and 325°K. Their results are shown graphically in Fig. 19. The theoretical values were computed

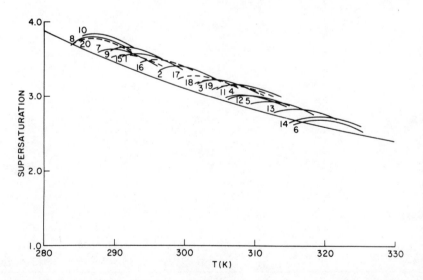

Fig. 19 Variation of the critical supersaturation of water vapor for homogeneous nucleation as a function of temperature with a diameter/height ratio of 6.8. The solid curve is the prediction of Becker–Döring theory. The envelope of the numbered curves represents the measured variation: – – –, H$_2$ as a carrier gas; ——, He as a carrier gas. From Heist and Reiss (1973) with permission of the American Institute of Physics.

from the classical steady-state theory of Becker and Döring using nucleation rates of 1 to 3 particles/cm^3-sec. The agreement between experiment and theory is quite good, the theoretical curve being about 8% below the experimental data.

Detailed rate measurements were made by Katz (1970) for n-hexane, n-heptane, n-octane, and n-nonane over the temperature range 225–330°K in a thermal diffusion chamber. He found that his data could be well explained by steady-state theory. Typical results and the comparison with theory are shown for the four gases in Figs. 20–23. In an earlier study Katz and Ostermier (1967) had examined ethanol, methanol, hexane, and H_2O in the temperature range 235–295°K. They obtained good agreement (usually within 5%) with the steady-state theory for ethanol, methanol, and hexane over a 60°K temperature range. For H_2O the data were limited to a 15° temperature range and the agreement between theory and experiment was within 15%. More recent work by Katz et al. (1975) on the n-alkyl-benzenes was also in good agreement with steady-state theory. Other studies of critical supersaturations have been made for various liquids using the expansion cloud chamber, the diffusion cloud chamber, and supersonic nozzles. These data have been collected and summarized by Pound (1972a).

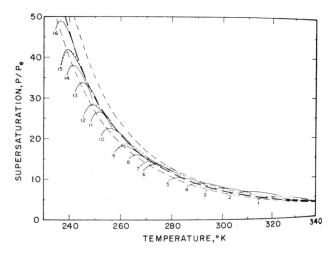

Fig. 20 Comparison of theory and experiment for the critical supersaturation of nonane in He needed to nucleate nonane drops versus temperature with a diameter:height = 7.44:1. The envelope to the numbered curves (not shown) is the experimental result. Dashed curves are the Becker–Döring steady-state theory calculations: — —, $J/\alpha = 1.0$; ———(upper), $J/\alpha = 100$; ———(lower), $J/\alpha = 0.01$. J is the nucleation rate and α is the accommodation coefficient. The dotted curves are for pressure doubled. From Katz (1970) with permission of the American Institute of Physics.

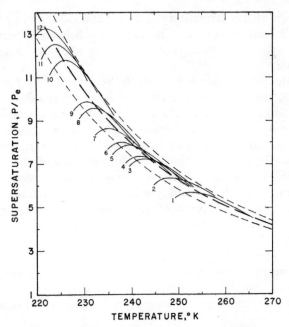

Fig. 21 Comparison of theory and experiment for the critical supersaturation of hexane in He needed to nucleate hexane drops versus temperature with diameter:height = 5.99:1. The envelope to the numbered curves (not shown) is the experimental result. Dashed curves are the Becker–Döring steady-state theory calculations: — —, $J/\alpha = 1.0$; ——— (upper), $J/\alpha = 100$; ——— (lower), $J/\alpha = 0.01$. J is the nucleation rate and α is the accommodation coefficient. From Katz (1970) with permission of the American Institute of Physics.

Nucleation from Chemical Reaction

$SO_2^*-C_2H_2$

Rate data exist for more complex chemical systems under isothermal conditions. For example, when SO_2 is photolyzed at 3000 to 3200 Å in the presence of hydrocarbons, particles are produced containing organic sulfur material. In the $SO_2-C_2H_2$ system, photolysis yields CO and particulate matter whose elemental analysis and molecular weight are consistent with the trimer of $C_3H_4S_2O_3$ (Luria et al., 1974a).

Typical particle growth curves for the particles are shown in Fig. 24. As the irradiation is started there is a short induction period followed by a burst of particle production due to homogeneous nucleation. Then particle production ceases and the ones already present grow mainly by condensation (there is some growth by coagulation) and decrease in number through

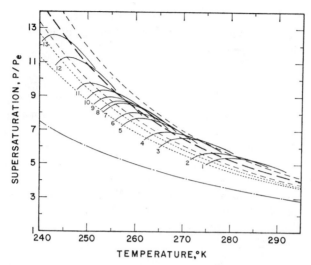

Fig. 22 Comparison of theory and experiment for the critical supersaturation of heptane in H_2 needed to nucleate heptane drops versus temperature with diameter:height = 5.99:1. The envelope to the numbered curves (not shown) is the experimental result. Curves are the Becker–Döring steady-state theory calculations: — —, $J/\alpha = 1.0$; ---(upper), $J/\alpha = 100$; ---(lower), $J/\alpha = 0.01$. The dotted line is the upper and lower estimate of the Reiss–Katz–Cohen theory with $J/\alpha = 1.0$ and –·– is the Lothe–Pound theory with $J/\alpha = 1.0$. J is the nucleation rate and α is the accommodation coefficient. From Katz (1970) with permission of the American Institute of Physics.

coagulation. Ultimately the particle number density is sufficiently reduced so that a second nucleation occurs, and there is a rapid increase in particle number density, as shown in two of the runs in Fig. 24.

The fundamental species C, in this case the trimer of $C_3H_4S_2O_3$, can undergo the reactions

$$C \to \text{wall removal}, \quad k_1$$
$$(q + 1)C \to C_{q+1}, \quad k_{\text{nucl}}$$

After particles are produced, condensation and coagulation become important as

$$C + C_n \to C_{n+1}, \quad k_{1,n}$$
$$C_n + C_m \to C_{n+m}, \quad k_{n,m}$$

The diffusion constant k_1 was measured directly by Luria et al. (1974a) to be 0.18 min^{-1}, and the overall average coagulation constant k_{coag} was obtained from second-order plots of particle density versus time to be $(1.5 \pm 0.2) \times 10^{-7}$ cm^3/min.

Once particles are produced, the concentration of C is in a steady state, so that its rate of formation (which in the experiment is $\frac{1}{3}$ the rate of CO production, $R\{CO\}/3$) equals its rate of removal

$$R\{CO\}/3 = k_{cond}N[C] + k_1[C] \tag{142}$$

where N is the particle number density and k_{cond} is the average value of the condensation coefficients $k_{1,n}$. In one experiment the time for the onset of nucleation was carefully measured so that the growth curve during the induction period could be used to measure [C] at the onset of nucleation. It was found to be 4.9×10^{10} molec/cm^3. Since the particle number density reaches its maximum very quickly, [C] does not change markedly between the onset of nucleation and the time for N to reach its maximum value N_{max}; N_{max} can be inserted for N in Eq. (142). Then Eq. (142) can be used to compute k_{cond} at N_{max} for each run.

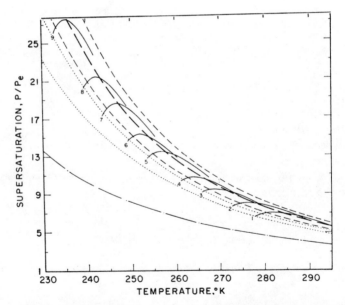

Fig. 23 Comparison of theory and experiment for the critical supersaturation of octane in H_2 needed to nucleate octane drops versus temperature with diameter:height = 7.44:1. The envelope to the numbered curves (not shown) is the experimental result. Curves are the Becker–Döring steady-state theory calculations: ——, $J/\alpha = 1.0$; ———(upper), $J/\alpha = 100$; ———(lower), $J/\alpha = 0.01$. The dotted line is the upper and lower estimate of the Reiss–Katz–Cohen theory with $J/\alpha = 1.0$ and —·— is the Lothe–Pound theory with $J/\alpha = 1.0$. J is the nucleation rate and α is the accommodation coefficient. From Katz (1970) with permission of the American Institute of Physics.

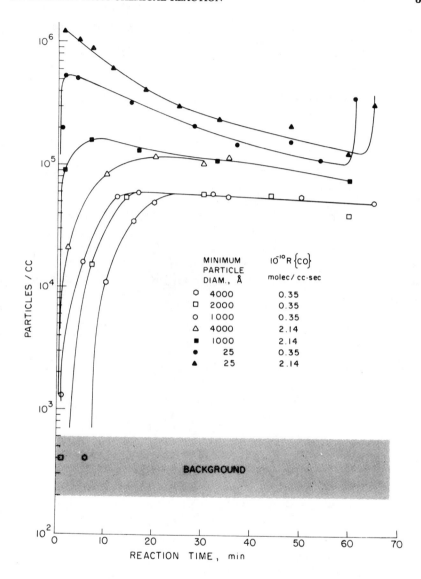

Fig. 24 Semilog plots of the particle density for particles greater than a specified minimum diameter versus reaction time for continuous radiation of a mixture containing 0.13 Torr SO_2, 1.9 Torr C_2H_2, and 1 atm N_2 at 29°C. From Luria et al. (1974a) with permission of the American Chemical Society.

At the maximum in particle density, the rate of particle production by nucleation equals the rate of particle removal by coagulation

$$k_{nucl}[C]^{q+1} = k_{coag}N_{max}^2 \qquad (143)$$

Since [C] can be obtained from Eq. (142), Eq. (143) can be rearranged as

$$\frac{R\{CO\}/3}{k_1 + k_{cond}N_{max}} = \left(\frac{k_{coag}}{k_{nucl}}\right)^{1/(q+1)} N_{max}^{2/(q+1)} \qquad (144)$$

In the SO_2–C_2H_2 system, $k_1 \ll k_{cond}N_{max}$, so that a further simplification can be effected as

$$\log N_{max} = \left(\frac{q+1}{q+3}\right)\log\left(\frac{R\{CO\}}{3}\right) - \frac{\log(k_{coag}k_{cond}^{q+1}/k_{nucl})}{q+3} \qquad (145)$$

A plot of $\log N_{max}$ vs $\log R\{CO\}$ is shown in Fig. 25. The plot is linear. From the slope $q + 1$ is found to be 3.6. From the intercept and the already determined values of k_{coag} and k_{cond}, k_{nucl} is found to be 3.4×10^{-30} cm$^{7.8}$/min.

SO_2^*–Allene

Very similar studies were made with SO_2–allene mixtures (Luria et al., 1974b). In this system the gaseous products were CO and C_2H_4. A liquid aerosol was produced tentatively identified as having the empirical formula C_5H_8SO. By an analysis analogous to that for SO_2–C_2H_2, the rate coefficients were evaluated, and they are listed in Table XIV.

Table XIV

Values of Rate Parameters for Nucleation by Chemical Reaction[a]

Rate parameter	$SO_2^* + C_2H_2$	$SO_2^* + C_3H_4$	$NH_3 + HNO_3$	$NH_3 + HCl$
Temp. (°C)	29	29	25	25
k_{diff} (min^{-1})	0.2	0.3	0.4[b]	
k_{coag} (cm^3/min)	1.5×10^{-7}	$\sim 10^{-7}$	1.3×10^{-7}	
k_{cond} (cm^3/min)	$(2-12) \times 10^{-7}$	7.0×10^{-7}	1.2×10^{-6} [c]	
k_{nucl} (cm^{3q}/min)	3.4×10^{-30}	3.0×10^{-59}	6.2×10^{-224} [d]	1.2×10^{-227} [e]
$q + 1$	3.6	6.7	8.0[f]	8.5[g]

[a] From Heicklen and Luria (1975).
[b] For HNO_3.
[c] $R_{cond} = k_{cond}[HNO_3]N$.
[d] cm^{45}/min.
[e] cm^{48}/min.
[f] $R_{nucl} = k_{nucl}[NH_3]^8[HNO_3]^8$.
[g] $R_{nucl} = k_{nucl}[NH_3]^{8.5}[HCl]^{8.5}$.

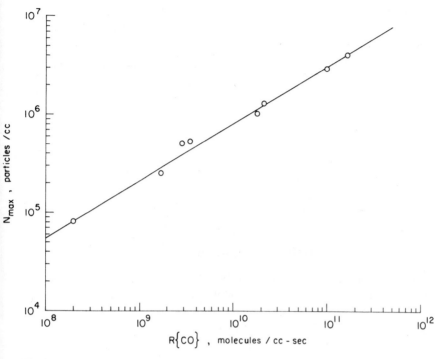

Fig. 25 Log–log plot of the maximum particle density versus the CO production rate for continuous radiation of mixtures of SO_2 and C_2H_2 in 1 atm of N_2 at 29°C. From Luria et al. (1974a) with permission of the American Chemical Society.

NH_3-HNO_3

The reaction of NH_3 with HNO_3 to produce NH_4NO_3 has also been studied (Olszyna et al., 1974). The HNO_3 was produced in the ozonlysis of NH_3 which proceeds by the overall stoichiometric reaction

$$NH_3 + 4O_3 \rightarrow 4O_2 + HNO_3 + H_2O$$

The rate law for the production of HNO_3 was found to be (Olszyna and Heicklen, 1972)

$$-R\{O_3\} = -d[O_3]/dt = k[O_3] \tag{146}$$

The particle curve growth curves are similar to those in Fig. 24, except that second nucleation was never observed.

The HNO_3 produced might react immediately with NH_3 to give monomeric NH_4NO_3 in the vapor phase. In that case, particle production would depend only on the rate of HNO_3 production, i.e., on $-R\{O_3\}/4$. On the other hand the monomeric NH_4NO_3 might be dissociated in the vapor phase, in which case particle production would also depend on the NH_3 concentration. Several experiments showed that particles were produced only when

$$-R\{O_3\}[NH_3]/4 = 2.3 \times 10^{27} \quad \text{molecules}^2/\text{cm}^6\text{-min}$$

Now the steady-state value for $[HNO_3]$ in runs in which particles are not produced is

$$[HNO_3] = -R\{O_3\}/4k_{diff} \tag{147}$$

where k_{diff} is the rate coefficient for HNO_3 removal on the walls

$$HNO_3 \rightarrow \text{wall removal}, \quad k_{diff}$$

The value for k_{diff} was estimated to be 0.4 min^{-1}. Since particle production was dependent on the product $[HNO_3][NH_3]$, it was concluded that NH_4NO_3 was dissociated in the vapor phase and that the vapor concentrations needed for particle production was

$$[NH_3][HNO_3] = 5.8 \times 10^{27} \quad \text{molecules}^2/\text{cm}^6$$

Rate coefficients for particle coagulation were obtained from second-order plots of the particle number density fall-off with reaction time after particle production had ceased. The value for k_{coag} was found to be 1.3×10^{-7} cm^3/min.

Particle growth by condensation could have occurred by two routes

$$HNO_3 + (NH_4NO_3)_n \rightleftarrows HNO_3 \cdot (NH_4NO_3)_n \quad (a, -a)$$

$$NH_3 + HNO_3 \cdot (NH_4NO_3)_n \rightarrow (NH_4NO_3)_{n+1} \tag{b}$$

or

$$NH_3 + HNO_3 \rightleftarrows NH_4NO_3 \quad (a', -a')$$

$$NH_4NO_3 + (NH_4NO_3)_n \rightarrow (NH_4NO_3)_{n+1} \tag{b'}$$

where n is an integer. These possibilities lead to three different limiting rate laws.

1. If reaction (a) is the rate-determining step, then

$$\exp\{k_{diff}\tau_{max}\} - 1 = k_{diff}/k_a N_{max} \tag{148}$$

2. If reaction (a') is the rate determining step, then

$$\exp\{k_{diff}\tau_{max}\} - 1 = k_{diff}/k_{a'}[NH_3] \tag{149}$$

3. If either reactions (b) or (b') are the rate-determining step, then

$$\exp\{k_{diff}\tau_{max}\} - 1 = k_{diff}/k''[NH_3]N_{max} \tag{150}$$

where N_{max} is the maximum value of the particle number density, τ_{max} is the time at N_{max}, and

$$k'' \equiv k_b k_a/k_{-a} + k_{b'} k_{a'}/k_{-a'}$$

The data indicated that the first rate law was operative and k_a was found to be 1.24×10^{-6} cm^3/min at τ_{max}.

Particle nucleation proceeds via the simplified equilibrium equation

$$(q + 1)NH_3 + (q + 1)HNO_3 \rightarrow (NH_4NO_3)_{q+1}$$

At N_{max} the rate of particle nucleation equals the rate of particle removal

$$k_{nucl}([NH_3][HNO_3])^{q+1} = k_{coag} N_{max}^2 \tag{151}$$

Now $[HNO_3]$ is given by its steady-state concentration

$$[HNO_3] = -R\{O_3\}/4(k_{diff} + k_a N_{max}) \tag{152}$$

so that

$$\frac{-R\{O_3\}[NH_3]/4}{1 + (k_a N_{max}/k_{diff})} = k_{diff}\left(\frac{k_{coag}}{k_{nucl}}\right)^{1/(q+1)} N_{max}^{2/(q+1)} \tag{153}$$

A log–log plot of the left-hand side of the equation versus N_{max} is shown in Fig. 26. From the slope and intercept it was found that

$$q + 1 = 8$$

$$k_{nucl} = 6.2 \times 10^{-224} \quad \text{cm}^{45}/\text{min}$$

It should be realized that even though two distinct vapors, HNO_3 and NH_3, are involved in the nucleation, it is still a homogeneous nucleation since particles are produced of a fixed composition.

NH_3-HCl

The nucleation rate for NH_4Cl formation from dry NH_3 and HCl vapors has been examined at room temperature by Twomey (1959) and Countess and Heicklen (1973). Both investigations utilized equal mixtures of NH_3 and HCl. A log–log plot of the nucleation rate versus reactant pressure is shown in Fig. 27. The lower pressure data are fitted with a straight line of slope 17. The higher pressure data give lower nucleation rates than pre-

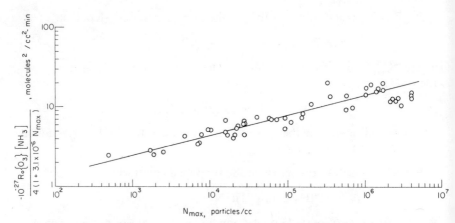

Fig. 26 Log–log plot of $(-R_0\{O_3\}[NH_3]/4)/(1 + 3.1 \times 10^{-6} N_{max})$ versus N_{max} for the reaction of NH_3 with O_3 in 760 Torr of N_2 at 25°C. From Olszyna et al. (1974) with permission of Pergamon Press.

dicted by the line. At the higher pressures the nucleation rates are large, and some coagulation may have occurred, thus reducing the particle count.

Twomey assumed that NH_4Cl was undissociated in the vapor phase. From none of the data can this hypothesis be ascertained, because the two reactant pressures were always the same. Accepting Twomey's assumption, however, leads to the rate law

$$dN/dt = k_{nucl}([NH_3][HCl])^{q+1} \tag{154}$$

From the data in Fig. 27, $q + 1 = 8.5$ and $k_{nucl} = 2 \times 10^{-229}$ cm^{48}/sec.

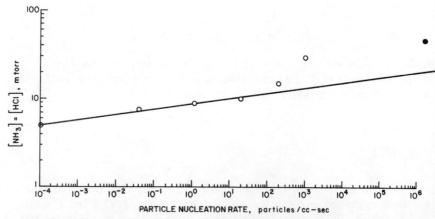

Fig. 27 Log–log plot of the particle production rate versus the NH_3 or HCl concentration in the reaction of NH_3 with HCl at room temperature. In each experiment the pressures of NH_3 and HCl were equal: ○, data of Twomey (1959); ●, data of Countess and Heicklen (1973). From Luria and Heicklen (1975), with permission of John Wiley and Sons.

CHAPTER VI

HETEROGENEOUS NUCLEATION

Homogeneous nucleation requires large supersaturations in order to form a stable nucleus capable of growth. For H_2O condensation in air, supersaturations >3 are needed at room temperature and below. For crystal precipitation from solution, much higher supersaturations are often required. If, however, some foreign particles are already present, then they can act as condensation nuclei and give rise to heterogeneous nucleation at lower supersaturations than needed for homogeneous nucleation. We shall consider separately the situation for soluble and insoluble foreign matter.

Soluble Condensation Nuclei

Let us consider the case of a vapor C condensing on a particle composed of molecules of A, the resulting particles having the formula $A_a \cdot C_n$, where

a and n are integers. We shall assume complete miscibility of A and C, and also that the solution is ideal, i.e.,

1. $\Delta H_{\text{soln}} = 0$, i.e., Raoult's law applies.
2. The volume of solution is the sum of the volumes of A_a and C_n.
3. The macroscopic surface tension $\bar{\gamma}_{a\cdot n}^\infty$ is the arithmetic average macroscopic surface tension, i.e.,

$$\bar{\gamma}_{a\cdot n}^\infty = (a\gamma_A + n\gamma_C)/(a + n) \tag{155}$$

The volume of A_a and C_n are, respectively,

$$V_a = (4\pi/3)r_A^3 a \tag{156}$$

$$V_n = (4\pi/3)r_C^3 n \tag{157}$$

where r_A and r_C are the molecular radii of A and C, respectively, as computed from macroscopic density measurements. Then the volume $V_{a\cdot n}$, area $A_{a\cdot n}$, and collision rate coefficient $Z_{1,a\cdot n}$ between C and $A_a \cdot C_n$ are, respectively,

$$V_{a\cdot n} = (4\pi/3)(ar_A^3 + nr_C^3) = (4\pi/3)r_{a\cdot n}^3 \tag{158}$$

$$A_{a\cdot n} = 4\pi(ar_A^3 + nr_C^3)^{2/3} = 4\pi r_{a\cdot n}^2 \tag{159}$$

$$Z_{1,a\cdot n} = \left(\frac{8\pi\kappa T}{\mu_{1,a\cdot n}}\right)^{1/2} (r_C + [ar_A^3 + nr_C^3]^{1/3})^2 \tag{160}$$

where $\mu_{1,a\cdot n} \equiv m_C(nm_C + am_A)/[m_C(n + 1) + am_A]$.

Nucleation

First let us consider an equilibrium theory for the rate of nucleation. As with homogeneous nucleation, we consider the reaction

$$A_a \cdot C_n + C \to A_a \cdot C_{n+1}, \quad k_{1,a\cdot n}$$

to be rate controlling and that $A_a \cdot C_n$ is in equilibrium with both vapors A and C

$$aA + nC \rightleftarrows A_a \cdot C_n \tag{161}$$

Then

$$[A_a \cdot C_n] = K_{\text{nucl}}[A]^a[C]^n \tag{162}$$

The rate of nucleation becomes

$$R_{\text{nucl}} = k_{1,a\cdot n} K_{\text{nucl}}[A]^a[C]^{n+1} \tag{163}$$

SOLUBLE CONDENSATION NUCLEI

The equilibrium constant K_{nucl} is directly obtained by realizing that reaction (161) can be considered to be the sum of three reactions

$$a\text{A} \rightleftarrows \text{A}_a \tag{164}$$

$$n\text{C} \rightleftarrows \text{C}_n \tag{165}$$

$$\text{A}_a + \text{C}_n \rightleftarrows \text{A}_a \cdot \text{C}_n \tag{166}$$

The equilibrium constants for reactions (164) and (165) are given by

$$K_{164} = \frac{1}{[\text{A}]_{\text{vp}}^{a-1}} \exp\left\{\frac{-4\pi r_\text{A}^2}{\kappa T} \gamma_a a^{2/3}\right\} \tag{167}$$

$$K_{165} = \frac{1}{[\text{C}]_{\text{vp}}^{n-1}} \exp\left\{\frac{-4\pi r_\text{C}^2}{\kappa T} \gamma_n n^{2/3}\right\} \tag{168}$$

The free energy for reaction (166) contains a surface energy term and a dilution term. There is no volume enthalpy of mixing because we have assumed an ideal solution. The respective free energy terms are

$$\Delta F_{a,n}^{\text{surf}} = 4\pi(r_{a \cdot n}^2 \gamma_{a \cdot n} - r_\text{A}^2 \gamma_a a^{2/3} - r_\text{C}^2 \gamma_n n^{2/3}) \tag{169}$$

$$\Delta F_{a,n}^{\text{dil}} = \kappa T\left(a \ln\left(\frac{a}{a+n}\right) + n \ln\left(\frac{n}{a+n}\right) + \ln\left[\frac{(a+n)[\text{A}]_{\text{vp}}[\text{C}]_{\text{vp}}}{a[\text{A}]_{\text{vp}} + n[\text{C}]_{\text{vp}}}\right]\right) \tag{170}$$

Since $K_{\text{nucl}} = K_{164} K_{165} K_{166}$, and

$$K_{166} = \exp\{(-\Delta F_{a \cdot n}^{\text{surf}} - \Delta F_{a \cdot n}^{\text{dil}})/\kappa T\} \tag{171}$$

$$K_{\text{nucl}} = \frac{(a+n)^{a+n-1} \exp\{-4\pi \gamma_{a \cdot n} r_{a \cdot n}^2/\kappa T\}(a[\text{A}]_{\text{vp}} + n[\text{C}]_{\text{vp}})}{[\text{A}]_{\text{vp}}^a [\text{C}]_{\text{vp}}^n a^a n^n} \tag{172}$$

The rate of nucleation from equilibrium theory becomes

$$R_{\text{nucl}} = \frac{Z_{1,a \cdot n}(a+n)^{a+n-1} \exp\{-4\pi \gamma_{a \cdot n} r_{a \cdot n}^2/\kappa T\}[\text{A}]^a [\text{C}]^{n+1}(a[\text{A}]_{\text{vp}} + n[\text{C}]_{\text{vp}})}{[\text{A}]_{\text{vp}}^a [\text{C}]_{\text{vp}}^n a^a n^n} \tag{173}$$

Now R_{nucl} is an overestimate of the rate regardless of the values of a and n adopted. Thus to obtain a least upper bound we take derivatives of R_{nucl} with respect to a and n and set them to zero. This, however, gives rise to a complex expression. Similarly the derivation of a steady-state rate law becomes extremely involved. Equation (173) can be extended to a multicomponent system in a straightforward manner.

Condensation

Now let us turn our attention to the reaction

$$A_a \cdot C_n + C \rightleftarrows A_a \cdot C_{n+1}$$

The equilibrium constant $K_{1,a \cdot n}$ can be obtained from the quotient of the equilibrium constants for the reactions

$$aA + (n+1)C \rightleftarrows A_a \cdot C_{n+1}$$

$$aA + nC \rightleftarrows A_a \cdot C_n$$

Thus

$$K_{1,a \cdot n} = \frac{(a[A]_{vp} + (n+1)[C]_{vp})(a+n+1)^{a+n} n^n}{(a[A]_{vp} + n[C]_{vp})(a+n)^{a+n-1}(n+1)^{n+1}[C]_{vp}}$$

$$\times \exp\left\{\frac{-4\pi}{\kappa T}(\gamma_{a \cdot n+1} r_{a \cdot n+1}^2 - \gamma_{a \cdot n} r_{a \cdot n}^2)\right\} \tag{174}$$

For C depositing on any particle $A_a \cdot C_{n+1}$, the rate coefficient is $k_{1,a \cdot n+1}$. The evaporation rate for the particle is $k_{-1,a \cdot n} = k_{1,a \cdot n}/K_{1,a \cdot n}$. Thus the effective rate coefficient for growth of any particle $A_a \cdot C_{n+1}$ by condensation of C, $k_{1,a \cdot n+1}\{\text{eff}\}$, is

$$k_{1,a \cdot n+1}\{\text{eff}\} = k_{1,a \cdot n+1}\left(1 - \frac{k_{1,a \cdot n}/K_{1,a \cdot n}}{k_{1,a \cdot n+1}[C]}\right) \tag{175}$$

At the steady state, $k_{1,a \cdot n+1}\{\text{eff}\} = 0$, $k_{1,a \cdot n}$ and $k_{1,a \cdot n+1}$ are given by their collision rate coefficients, and the steady-state supersaturation $S_{1,a \cdot n+1}$ is

$$S_{1,a \cdot n+1} = \frac{Z_{1,a \cdot n}(a[A]_{vp} + n[C]_{vp})(a+n)^{a+n-1}(n+1)^{n+1}}{Z_{1,a \cdot n+1}(a[A]_{vp} + (n+1)[C]_{vp})(a+n+1)^{a+n} n^n}$$

$$\times \exp\left\{\frac{4\pi}{\kappa T}(\gamma_{a \cdot n+1} r_{a \cdot n+1}^2 - \gamma_{a \cdot n} r_{a \cdot n}^2)\right\} \tag{176}$$

These expressions for K_{nucl}, R_{nucl}, $K_{1,a \cdot n}$, and $S_{1,a \cdot n+1}$ are the most general forms. If the equilibrium vapor pressures, surface tensions, and molar densities are known, they can be evaluated for any a and n.

Usually, however, we are interested in a more specific situation which leads to simple solutions, i.e., when

$$[A]_{vp} = 0, \qquad a, n \gg 1$$

In this situation we start with a particle of fixed size A_a and condense C on it. The general reaction of interest is

$$A_a \cdot C_n + C \rightleftarrows A_a \cdot C_{n+1}$$

where a is now a fixed number. The equilibrium of interest becomes

$$A_a + nC \rightleftarrows A_a \cdot C_n$$

which can be considered as the product of two equilibria

$$A_a + C \rightleftarrows A_a \cdot C$$

$$A_a \cdot C + (n-1)C \rightleftarrows A_a \cdot C_n$$

Thus K_{nucl} becomes

$$K_{\text{nucl}} = \frac{(a+n)^{a+n-1} K_{1,a\cdot 0}}{n^{n-1}(a+1)^a [C]_{\text{vp}}^{n-1}} \exp\{-4\pi(\gamma_{a\cdot n} r_{a\cdot n}^2 - \gamma_{a\cdot 1} r_{a\cdot 1}^2)/\kappa T\} \quad (177)$$

and the nucleation rate is

$$R_{\text{nucl}} = Z_{1,a\cdot n} K_{\text{nucl}} [A_a][C]^{n+1} \quad (178)$$

To evaluate $K_{1,a\cdot n}$ and $S_{1,a\cdot n+1}$, we realize that $Z_{1,a\cdot n}/Z_{1,a\cdot n+1} \sim 1$, and that the exponential term can be expanded by the binomial theorem. The simplified equations obtained are

$$K_{1,a\cdot n} = \frac{a+n}{n[C]_{\text{vp}}} \exp\left\{\frac{-8\pi r_C^3 \bar{\gamma}_{a\cdot n}^\infty}{3\kappa T(a+n)^{1/3} \bar{r}_1}\right\} \quad (179)$$

and

$$S_{1,a\cdot n+1} = \frac{n}{a+n} \exp\left\{\frac{8\pi r_C^3 \bar{\gamma}_{a\cdot n}^\infty}{3\kappa T(a+n)^{1/3} \bar{r}_1}\right\} \quad (180)$$

where we have introduced the average value $\bar{r}_1$ defined by

$$\bar{r}_1 \equiv (ar_A^3 + nr_C^3)^{1/3}/(a+n)^{1/3}$$

Both the exact and approximate forms of $S_{1,a\cdot n+1}$ go through a maximum as n increases. Thus if C is placed in the presence of A_a, it will start to condense until the steady-state supersaturation is achieved, and then condensation will stop. The particles can only grow indefinitely if $[S_1]$ exceeds the maximum value of $S_{1,a\cdot n+1}$ for all n. The value of n at the maximum can be found by taking the derivative of $S_{1,a\cdot n+1}$ with respect to n and setting it to zero. It is easiest to take the derivative if we realize that $\bar{r}_1$ and $\bar{\gamma}_{a\cdot n}^\infty$ are very slowly varying functions, especially if a and n have quite different values. Thus if we assume these functions to be constants the derivative yields a maximum value for n, n_{max}, at

$$a + n = (8\pi n r_C^3 \bar{\gamma}_{a \cdot n}^\infty / 9\kappa T a \bar{r}_1)^3 \tag{181}$$

For any reasonable values of the parameters, $n \gg a$, so that

$$n \sim (\bar{r}_1 9\kappa T a / 8\pi r_C^3 \bar{\gamma}_{a \cdot n}^\infty)^{3/2} \tag{182}$$

Then the maximum value for the supersaturation $S_{1,a \cdot n+1}^{max}$, becomes

$$S_{1,a \cdot n+1}^{max} \simeq \exp\left\{\frac{(8\pi r_C^3 \bar{\gamma}_{a \cdot n}^\infty / 3\kappa T)^{3/2}}{(3a)^{1/2} \bar{r}_1^{3/2}}\right\} \tag{183}$$

Furthermore since $n \gg a$, $\bar{r}_1^2 \sim r_C^2$ and $\bar{\gamma}_{a \cdot n}^\infty \sim \gamma_C$.

Table XV shows the values of $S_{1,a \cdot n+1}$ computed from Eq. (176) for H_2O at 25°C ($4\pi r_C^2 \gamma_\infty / \kappa T = 8.1693$) assuming that A_a has no vapor pressure and that $\gamma_A = \gamma_C$, $r_A = r_C$, and $m_A = m_C$. The entries are calculated for $\gamma_{a \cdot n} = \gamma_\infty [1 - (r_C/r_{a \cdot n})^3]$. With 10 A molecules, $S_{1,a \cdot n+1}^{max}$ is 3.84, which corresponds to the value needed for homogeneous nucleation at 25°C. The maximum value quickly drops to 1.57 for $a = 100$, 1.16 for $a = 10^3$, 1.05 for $a = 10^4$, and 1.004 for $a = 10^5$. These trends are seen readily in Fig. 28, which is a plot of $S_{1,a \cdot n+1}$ vs $a + n$, so that the effect of dilution can be compared directly. Also plotted in Fig. 28 are curves when $[A]_{vp} = [C]_{vp}$, i.e. the two species are indistinguishable in terms of their vapor pressures. It is apparent, that the vapor pressure of A has only a very slight effect on $S_{1,a \cdot n+1}$. Essentially all the reduction in $S_{1,a \cdot n+1}$ from the homogeneous case ($a = 0$) is due to the entropy of mixing.

It should be realized that, contrary to the case of homogeneous nucleation where any particle tends to grow or shrink, in heterogeneous nucleation particles with $n < n_{max}$ are extremely stable when in contact with vapor at their steady state supersaturation. This occurs because the slope of the $S_{1,a \cdot n+1}$ vs n curve is positive rather than negative, so that a slight growth favors vaporization, whereas a slight contraction promotes growth.

These developments assumed ideal behavior. If nonideality occurs, the treatment is the same except that the correct volume, surface tension, and vapor pressures for each composition must be used. Also if the solute dissociates in solution, the van't Hoff factor must be introduced, since each fragment then behaves separately.

Insoluble Condensation Nuclei

We consider the case now of insoluble nuclei. If the insoluble material is not wetted by the condensing vapor, i.e., there is no heat of adhesion,

Table XV

Values of $S_{1,a\cdot n+1}$ for $4\pi r_C^2 \gamma_\infty/\kappa T = 8.1693$ [a]

n	a = 10	10^2	10^3	10^4	10^5	10^6
1	8.349 − 01	2.354 − 02	1.267 − 03	9.473 − 05	8.274 − 06	7.769 − 07
2	1.592	5.224 − 02	2.847 − 03	2.131 − 04	1.861 − 05	1.748 − 06
4	2.597	1.103 − 01	6.165 − 03	4.624 − 04	4.040 − 05	3.793 − 06
6	3.162	1.665 − 01	9.529 − 03	7.162 − 04	6.258 − 05	5.877 − 06
8	3.485	2.201 − 01	1.290 − 02	9.715 − 04	8.490 − 05	7.974 − 06
10	3.669	2.711 − 01	1.626 − 02	1.227 − 03	1.073 − 04	1.007 − 05
12	3.772	3.197 − 01	1.961 − 02	1.483 − 03	1.297 − 04	1.218 − 05
14	3.824	3.660 − 01	2.295 − 02	1.740 − 03	1.521 − 04	1.429 − 05
16	3.843	4.100 − 01	2.627 − 02	1.996 − 03	1.746 − 04	1.639 − 05
18	3.842	4.520 − 01	2.958 − 02	2.252 − 03	1.970 − 04	1.850 − 05
20	3.828	4.921 − 01	3.288 − 02	2.508 − 03	2.195 − 04	2.061 − 05
25	3.760	5.844 − 01	4.106 − 02	3.148 − 03	2.756 − 04	2.589 − 05
30	3.674	6.668 − 01	4.914 − 02	3.788 − 03	3.318 − 04	3.117 − 05
40	3.498	8.073 − 01	6.504 − 02	5.066 − 03	4.441 − 04	4.172 − 05
70	3.090	1.095	1.106 − 01	8.882 − 03	7.810 − 04	7.339 − 05
100	2.827	1.264	1.533 − 01	1.267 − 02	1.118 − 03	1.051 − 04
300	2.162	1.569	3.795 − 01	3.735 − 02	3.357 − 03	3.161 − 04
1×10^3	1.703	1.540	7.700 − 01	1.161 − 01	1.112 − 02	1.054 − 03
3×10^3	1.453	1.406	1.057	2.909 − 01	3.271 − 02	3.158 − 03
1×10^4	1.286	1.274	1.161	6.111 − 01	1.018 − 01	1.045 − 02
3×10^4	1.191	1.187	1.151	8.795 − 01	2.570 − 01	3.074 − 02
1×10^5	1.124	1.123	1.113	1.019	5.488 − 01	9.583 − 02
3×10^5	1.085	1.084	1.081	1.049	8.075 − 01	2.426 − 01
1×10^6	1.056	1.056	1.055	1.045	9.583 − 01	5.221 − 01
3×10^6	1.038	1.038	1.038	1.035	1.004	7.753 − 01
1×10^7	1.026	1.026	1.026	1.025	1.015	9.319 − 01
3×10^7	1.018	1.018	1.018	1.017	1.014	9.849 − 01
1×10^8	1.012	1.012	1.012	1.012	1.012	1.002

[a] Entries give the value and the power of 10; e.g., $1.234 - 06 = 1.234 \times 10^{-6}$. $\gamma_A = \gamma_C$, $r_A = r_C$, $m_A = m_C$, $\gamma_{a\cdot n} = \gamma_\infty[1 - (r_C/r_{a\cdot n})^3]$. Assumes A_a has no vapor pressure and A_a forms an ideal solution with C_n.

then the nucleus will not aid condensation. If, however, the nucleus is wetted, i.e., there is a heat of adhesion, then condensation can be promoted.

Consider a nucleus A_a which increases the enthalpy of vaporization H_{vap} of a molecule of C in the surface of C_n by an inverse power law of the distance

$$\Delta H_{vap} = \lambda/r_{a\cdot n}^v \tag{184}$$

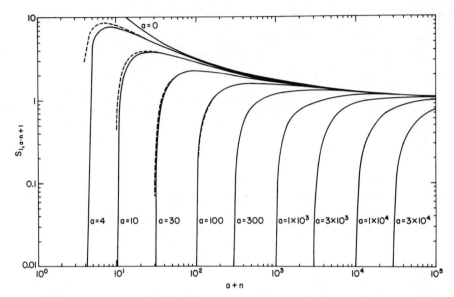

Fig. 28 Plot of the steady-state supersaturation of C with $A_a \cdot C_{n+1}$, $S_{1,a \cdot n+1}$, versus $a + n$ for various values of a with $m_A = m_C$, $\gamma_A = \gamma_C$, $r_A = r_C$, and $\gamma_{a \cdot n} = \gamma_C[1 - (r_C/r_{a \cdot n})^3]$. The solid line is vapor pressure of A equal to zero; the broken line is vapor pressure of A and C equal. It is assumed that A and C form an ideal undissociated solution at all concentrations.

Many systems show such dependencies, such as

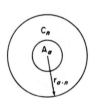

Interaction	v
charge–charge	1
charge–dipole	2
dipole–dipole	3
charge–induced dipole	4
dipole–induced dipole	5
induced dipole–induced dipole	6

A pictorial representation of our growing drop is illustrated above. That the interaction energy ΔH_{vap} between the nucleus and a surface molecule in $A_a \cdot C_{n+1}$ will follow Eq. (184) is based on two assumptions:

1. A_a is located in the center of the drop.
2. The radius of A_a is small compared to $r_{a \cdot n}$.

The first assumption becomes more and more valid as A_a increases in size for $v \geq 3$. For a small nucleus A_a, it can wander in the larger condensate drop, the smaller it is the more random its position. The center of the large drop will, however, be the most preferred position for $v \geq 3$ because this position minimizes the total interaction energy. For $v = 1$, i.e., for a Coulombic interaction, the interaction energy between A_a and all the molecules in C_{n+1} is independent of the location of A_a; each position in phase space will be equally likely, and the nucleus, which in this case is an ion, will most often be in the surface. (A "dissolved" ion is insoluble in the sense that the total drop is not homogeneous.) For $v = 2$, the most probable position will be inside the drop, but not at the center. Thus our assumption 1 is really good only if an ion is not involved (i.e., for $v \geq 3$). In this section, we will not include ions as condensation nuclei; they will be discussed separately in the next section.

The second assumption comes from the fact that we have assumed the interaction potential to depend only on the distance between a surface molecule C and the center of A_a. Actually the interaction occurs to every point in A_a, and a summation is needed. Only for $v = 1$ will the interaction potential depend only on the distance between centers.

Actually assumptions 1 and 2 are, to a certain extent, mutually exclusive. Nevertheless, over a short change in distance, the interaction energy ΔH_{vap} can be approximated by Eq. (184). The radius of $A_a \cdot C_n$ is given by

$$r_{a\cdot n} = [ar_A^3 + nr_C^3]^{1/3} \tag{185}$$

and the ratio of the steady-state supersaturation $S_{1,a\cdot n+1}$ to that for a drop of pure C with the same radius as $A_a \cdot C_{n+1}$, $S^\circ_{1,a\cdot n+1}$, is

$$S_{1,a\cdot n+1} = S^\circ_{1,a\cdot n+1} \exp\left\{\frac{-\lambda/\kappa T}{[ar_A^3 + (n+1)r_C^3]^{v/3}}\right\} \tag{186}$$

Since the surface of the drop is entirely composed of C, the steady-state supersaturation becomes

$$S_{1,a\cdot n+1} = \frac{Z_{1,a\cdot n}}{Z_{1,a\cdot n+1}} \exp\left\{\frac{-\lambda/\kappa T}{[ar_A^3 + (n+1)r_C^3]^{v/3}}\right\}$$
$$\times \exp\left\{\frac{4\pi\gamma_\infty}{\kappa T}([r_A^3 a + r_C^3(n+1)]^{2/3} - [r_A^3 a + r_C^3 n]^{2/3})\right\} \tag{187}$$

where γ_∞ is the surface tension of pure C_n, and it has been assumed that the drop is large enough so that the macroscopic surface tension is applicable.

For condensation to be favored, it is necessary that the supersaturation exceed the maximum value of $S_{1,a\cdot n+1}$ for a given a. It is easier to find this maximum for the approximation: $a + n \gg 1$. For this condition, $Z_{1,a\cdot n}/Z_{1,a\cdot n+1} \sim 1$, and

$$S_{1,a\cdot n+1} \simeq \exp\left\{\frac{8\pi r_C^3 \gamma_\infty}{3\kappa T(a+n)^{1/3}\bar{r}_1} - \frac{v\lambda r_C^3/3\kappa T}{\bar{r}_1^{v+3}(a+n)^{1+v/3}}\right\} \tag{188}$$

where

$$\bar{r}_1 \equiv (ar_A^3 + nr_C^3)^{1/3}/(a+n)^{1/3}$$

If the adsorption energy is negligibly small, then $\lambda \Rightarrow 0$, and $S_{1,a\cdot n+1}$ is a maximum for $n = 0$

$$S_{1,a\cdot n+1}^{\max} = \exp\left\{\frac{8\pi r_C^3 \gamma_\infty}{3\kappa T a^{1/3} r_A}\right\}, \quad \lambda \sim 0 \tag{189}$$

If λ is significantly large, then the maximum value of $S_{1,a\cdot n+1}$ is found by assuming that $\bar{r}_1$ is constant, taking the derivative of $S_{1,a\cdot n+1}$ with respect to n, and setting it to zero. Then

$$(n_{\max} + a) = \left[\frac{(3+v)\lambda v}{8\pi \gamma_\infty \bar{r}_1^{v+2}}\right]^{3/(2+v)} \tag{190}$$

and

$$S_{1,a\cdot n+1}^{\max} = \exp\left\{\frac{8\pi r_C^3 \gamma_\infty}{3\kappa T}\left(\frac{8\pi \gamma_\infty}{(3+v)v\lambda}\right)^{1/(2+v)} - \frac{vr_C^3}{3\kappa T}\left[\frac{8\pi \gamma_\infty}{(3+v)v\lambda}\right]^{(3+v)/(2+v)}\right\} \tag{191}$$

The rate of condensation can be obtained from considering the overall reaction

$$A_a + nC \leftrightarrows A_a \cdot C_n$$

The equilibrium constant K_{nucl} is

$$K_{\text{nucl}} = \frac{K_{1,a\cdot 0}}{[C]_{vp}^{n-1}} \exp\left\{\frac{-4\pi}{\kappa T}(\gamma_{a\cdot n} r_{a\cdot n}^2 - \gamma_{a\cdot 1} r_{a\cdot 1}^2) + \frac{\lambda}{\kappa T}\sum_{j=2}^{n}\frac{1}{r_{a\cdot j}^v}\right\} \tag{192}$$

and the equilibrium theory nucleation rate is

$$R_{\text{nucl}} = Z_{1,n} K_{1,a\cdot 0} S_1^{n+1}[A_a][C]_{vp}$$
$$\times \exp\left\{\frac{-4\pi}{\kappa T}(\gamma_{a\cdot n} r_{a\cdot n}^2 - \gamma_a r_a^2) + \frac{\lambda}{\kappa T}\sum_{j=2}^{n}\frac{1}{r_{a\cdot j}^v}\right\} \tag{193}$$

As before, R_{nucl} calculated from Eq. (193) is an upper limit, and the least upper bound is obtained by minimizing with respect to n.

The steady-state theory solution for R_{nucl} becomes

$$R_{\text{nucl}} = \frac{k_{1,n} K_{1,a\cdot 0} S_1^{n+1}[A_a][C]_{vp}^2 \exp\{(-4\pi/\kappa T)(r_{a\cdot n}^2 \gamma_{a\cdot n}^2 - \gamma_{a\cdot 1} r_{a\cdot 1}^2) + (\lambda/\kappa T)\sum_{i=2}^{n}(1/r_{a\cdot i}^v)\}}{1 + k_{1,n}\sum_{j=1}^{n-1}(S_1^j/k_{1,n-j})\exp\{(-4\pi/\kappa T)(\gamma_{a\cdot n} r_{a\cdot n}^2 - \gamma_{a\cdot n-j+1} r_{a\cdot n-j+1}^2) + (\lambda/\kappa T)\sum_{i=n-j+2}^{n}(1/r_{a\cdot i}^v)\}}$$

(194)

where we have assumed that $A_a \cdot C$ is in equilibrium with $A_a + C$.

Ions as Condensation Nuclei

For ions as condensation nuclei, $v = 2$ and the interaction potential parameter λ becomes

$$\lambda = e(\mathbf{q} + e\alpha/2r_{a\cdot n+1}^2)/\varepsilon_{a\cdot n+1} \tag{195}$$

where $\mathbf{q}$ is the dipole moment, α the molecular polarizability, e the electronic change, and $\varepsilon_{a\cdot n+1}$ the dielectric constant of $A_a \cdot C_{n+1}$. (A_a is now the ion.) The increase in H_{vap} becomes

$$\Delta H_{\text{vap}} = e(\mathbf{q} + e\alpha/2r_{a\cdot n+1}^2)/\varepsilon_{a\cdot n+1} r_{a\cdot n+1}^2 \tag{196}$$

The induced dipole is $\alpha/r_{a\cdot n+1}^2$, but its contribution to the energy of interaction is only $\frac{1}{2}$ that of a permanent dipole. Equation (196) is applicable as long as the charge is distributed spherically symmetrically in the drop.

For very large particles Eq. (196) is easy to evaluate since $\varepsilon_{a\cdot n+1} \Rightarrow \varepsilon_\infty$, the macroscopic dielectric constant of C. As n decreases, however, $\varepsilon_{a\cdot n+1}$ deviates from its macroscopic value because of the importance of surface effects and deviations near the ion. Thus the problem we must face is how to evaluate $\varepsilon_{a\cdot n+1}$ for small n.

For an ion, as discussed in the previous section, the charge tends to concentrate in the surface of the drop. In a perfectly conducting sphere, the charge will be evenly spaced on the surface, and for an electronic charge, Thompson in 1888 showed the potential energy to be

$$e^2/2r_{a\cdot n}$$

If the sphere is not perfectly conducting, Volmer (1939) introduced the correction term

$$(1 - \varepsilon_0/\varepsilon_\infty)$$

where ε_0 is the dielectric constant of the surrounding medium (i.e., the vapor). Thus the potential energy would be

$$(e^2/2r_{a\cdot n})(1 - \varepsilon_0/\varepsilon_\infty)$$

This form gives the correct limiting rate law when $\varepsilon_\infty \Rightarrow \infty$, but leads to two incorrect predictions:

(1) It predicts the interaction energy to be zero when $\varepsilon_\infty = \varepsilon_0$.

(2) It predicts a decrease in the potential energy of the same proportion regardless of the size of the drop.

The first prediction is wrong since it infers that the energy of interaction depends on the surrounding medium. The second prediction is incorrect since, in fact, for a small particle, where the interaction energy is large, the charge is going to distribute itself over the surface, in spite of the fact that the sphere is not perfectly conducting, i.e., charge will move through a dielectric if the breakdown potential is exceeded.

Consequently the Thompson energy should be applicable for small drops of finite dielectric constant as well as for large drops of infinite dielectric constant. For this interaction energy, ΔH_{vap} is given by

$$\Delta H_{vap} = \frac{e^2}{2}\left[\frac{1}{r_{a \cdot n}} - \frac{1}{r_{a \cdot n+1}}\right] \tag{197}$$

The dielectric constant $\varepsilon_{a \cdot n+1}$, must behave in such a way that both Eqs. (196) and (197) are satisfied under the appropriate conditions. A function which satisfies this criterion is

$$\varepsilon_{a \cdot n+1} = \varepsilon_\infty(1 - \exp\{-\beta_{a \cdot n+1}/\varepsilon_\infty\}) \tag{198}$$

For $\varepsilon_\infty \Rightarrow \infty$, then $\varepsilon_{a \cdot n+1} = \beta_{a \cdot n+1}$, so that $\beta_{a \cdot n+1}$ is given by

$$\beta_{a \cdot n+1} = \frac{2[(\mathbf{q}/e) + (\alpha/2r_{a \cdot n+1}^2)]}{r_{a \cdot n+1}^2\left[\dfrac{1}{r_{a \cdot n}} - \dfrac{1}{r_{a \cdot n+1}}\right]} \tag{199}$$

Now Eq. (196) can be used as a general expression by evaluating $\varepsilon_{a \cdot n+1}$ from Eq. (198).

The steady-state supersaturation $S_{1,a \cdot n+1}$, now must incorporate the extra energy term ΔH_{vap} due to the ion, which leads to

$$S_{1,a \cdot n+1} = \frac{Z_{a \cdot n}}{Z_{a \cdot n+1}} \exp\left\{\frac{-e^2[(\mathbf{q}/e) + (\alpha/2r_{a \cdot n+1}^2)]}{\kappa T r_{a \cdot n+1}\varepsilon_{a \cdot n+1}}\right\}$$

$$\times \exp\left\{\frac{4\pi}{\kappa T}[\gamma_{a \cdot n+1}r_{a \cdot n+1}^2 - \gamma_{a \cdot n}r_{a \cdot n}^2]\right\} \tag{200}$$

Comparison with Experiment

Wilson (1897, 1900), in his famous cloud chamber experiments at the end of the 19th century discovered that H_2O condensed sooner in the presence of ions and that negative ions were more effective than positive ions. In these experiments H_2O at its vapor pressure (in air) at room temperature was adiabatically expanded to reduce the temperature. Thus the vapor became supersaturated (even though the absolute pressure dropped). The greater the expansion, the greater the temperature reduction, and thus the greater the supersaturation. By measuring the expansion ratio needed for condensation to appear, the final temperature and supersaturation could be computed.

These investigations were continued by other workers, both with H_2O and other vapors. The results of the H_2O experiments are tabulated in Table XVI. For negative ions the temperature and supersaturation observed are 265–268°K and 3.7–4.2, respectively. At 268°K, homogeneous nucleation requires a critical supersaturation of 4.55. For positive ions, further expansion was required to produce supersaturations of ~ 6.0 for condensation. The investigators did not list the temperatures, but they must be slightly less than 265°K.

For other vapors it was sometimes found that positive ions were more effective than negative ions, and in some cases it made no difference. This phenomenon was investigated carefully by Loeb et al. (1938). Their results, with the expansion E needed for condensation, are shown in Table XVII.

The theory of condensation just derived does not differentiate between positive and negative ions. In order for the charge to be distributed on the outside of the sphere, however, the molecules in the liquid drop must align their dipoles so that the appropriately charged end is outward. Thus for a positive ion, the positive end of the dipole must be at the surface, whereas for a negative ion the reverse is true. In the two different configurations, the surface free energy may be different, and thus the supersaturation required for condensation will be different. For molecules with no dipole moment, positive and negative ions should have the same effect. For the two such species tested, C_6H_6 and CCl_4, this has been found to be the case.

White and Kassner (1971) pointed out that neutral and negatively charged water both have their dipoles pointing inward, i.e., the surface is negatively

Table XVI

Supersaturation at which Condensation of H_2O Vapor Occurs on Small Ions in Air

Temp. (°K)	Supersaturation	Reference
	Positive ions	
	6.0	Wilson (1900)
	6.0	Przibram (1906)
	4.87	Scharrer (1939)
	Negative ions	
267.8	4.2	Wilson (1900)
	3.7	Przibram (1906)
267.6	4.2	Laby (1908)
	4.1	Andrén (1917)
266.5	3.98	Powell (1928)
265	4.1	Flood (1934)
	4.14	Scharrer (1939)
265	3.9	Sander and Damköhler (1943)

Table XVII

Expansion Ratios Needed for Nucleation on Ions for Saturated Vapors Initially at Room Temperature[a]

Compound	E_-	E_+	$E_0{}^b$
	No ion preference		
$C_6H_5NO_2$	1.45	1.45	—
$CHCl_3$	1.83	1.83	—
Acetone	1.88	1.88	2.06
C_6H_6	2.15	2.15	—
	Negative ion preferred		
H_2O	1.25	1.31	1.32
Aniline	1.44	1.48	—
C_6H_5Cl	1.53	1.6	1.8
Toluene	1.60	1.69	1.73
C_2H_5I	1.82	2.11	—
	Positive ion preferred		
C_2H_5OH	1.29	1.26	—
CH_3OH	1.37	1.33	—
CH_3COOH	1.54	1.45	1.58
$n\text{-}C_4H_9Br$	—	1.48	—

[a] From Loeb et al. (1938).
[b] E_0 refers to ions absent.

charged. Thus negative ions retain the surface energy of uncharged water and give the maximum reduction in free energy. With positively charged ions, the dipoles are reversed, and the drop has an additional energy term. They gave the energy difference per surface molecule ΔE as

$$\Delta E = \frac{75}{16}\frac{qQ_{zz}}{r_C^2}\left(\frac{1}{15} + \frac{\varepsilon_{a\cdot n}}{(1 + 2\varepsilon_{a\cdot n})(2 + 3\varepsilon_{a\cdot n})}\right) \quad (201)$$

where Q_{zz} are elements of the diagonalized quadrupole tensor, and $\varepsilon_{a\cdot n}$ the effective dielectric constant of the cluster. They measured the number of drops produced in a cloud chamber experiment starting with saturated H_2O vapor at 25°C. Their results are shown in Fig. 29. The theoretical rates were calculated using Frenkel's (1939) procedure. The theory satisfactorily fits the experimental points.

The quantitative results for the supersaturations and temperatures required for supersaturation are given in Table XVIII. Both calculated and observed values for the critical supersaturations are listed. The calculated values are the maximum values $S_{1,a\cdot n+1}^{\max}$ computed from Eq. (200). Listed

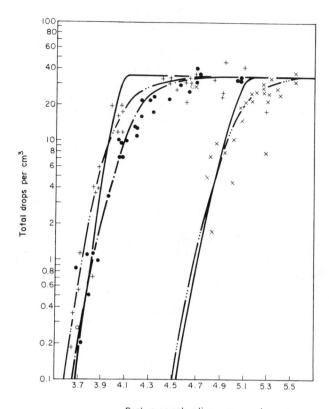

Fig. 29 Number of drops of H_2O nucleated as a function of peak supersaturation in the adiabatic expansion of water vapor initially at 25°C at its vapor pressure in He or Ar. The solid line is for the theoretical values. The points are: ●, helium(−); +, argon(−); and ×, argon(+). From White and Kassner (1971) with permission of Pergamon Press.

along with them are the values of n at the maximum n_{max}. The other parameters in the table were taken from "The Handbooks of Physics and Chemistry," interpolated or extrapolated from data in the handbooks, or from the values given by Laby (1908). The polarizabilities α were calculated from

$$\alpha = \frac{(\varepsilon_\infty - 1)}{(\varepsilon_\infty + 2)} r_C^3 \qquad (202)$$

Since the specific ion involved was not given, it was assumed arbitrarily that $r_A = r_C$ and $m_A = m_C$ in making the calculations. Also the value of γ_∞ used is for the neutral species, and is not quite exact for charged droplets.

Table XVIII
Supersaturation at Which Condensation Occurs on Small Ions for Various Liquids

Compound	Ion sign	Temp. (°K)	ρ (g/cm^3)	$10^8 r_C$ (cm)	γ_∞ (dynes/cm)	$10^8 q,e$ (cm)	ϵ_∞	$10^{24}\alpha$ (cm^3)	n_{max}^a	$S_{1,a,n+1}^{max}$ {calc}^{a}	S_1 {obs}	Reference for S_1 {obs}
						Esters						
Ethyl Acetate	+	247.2	0.956	3.318	31.8	0.373	5.94	22.72	19	8.84	8.9	1
Methyl Butyrate	+	258.2	0.935	3.513	28.2	–	–	–	–	–	5.3	1
Methyl Isobutyrate	+	259.2	0.931	3.518	26.8	–	–	–	–	–	5.2	1
Propyl Acetate	+	260.1	0.928	3.522	27.3	–	–	–	–	–	5.0	1
Ethyl Propionate	+	252.5	0.94	3.507	28.1	–	–	–	–	–	7.8	1
						Acids						
HCOOH	+	230.6	1.29	2.418	43.8	0.294	~70	13.55	30	5.0	25.1	1
CH$_3$COOH	+	251.1	1.09	2.848	25.9	0.362	5.6	13.98	50	2.4	9.3	1
C$_2$H$_5$COOH	+	257.5	1.03	3.055	29.6	0.364	3.18	11.99	40	3.0	9.4	1
n-C$_3$H$_7$COOH	+	254.7	0.997	3.272	29.4	–	3.06	14.26	–	–	15.0	1
i-C$_3$H$_7$COOH	+	256.2	0.987	3.283	27.5	–	2.66	12.60	–	–	13.3	1
Isovaleric Acid	+	267.7	–	–	–	–	–	–	–	–	6.0	1
						Alcohols						
CH$_3$OH	+	266.3	0.826	2.486	25.4	0.354	33.56	14.07	50	2.0	3.1	2
	+	264	–	–	–	–	–	–	–	–	1.85	6
	+	270	–	–	–	–	–	–	–	–	1.8	6
	+	276	–	–	–	–	–	–	–	–	1.75	6
	+	264.5	0.82	2.492	25.4	0.354	33.56	14.18	–	–	2.95	3
	–	262	0.82	2.492	25.4	0.354	33.56	14.18	–	–	3.39	3

Compound	±												Ref.
C$_2$H$_5$OH	+	272.7	0.81	2.825	23.8	0.352	24.2	19.96	35	2.5	2.3		2
	+	264	–	–	–	–	–	–	–	–	2.2		6
	+	269	–	–	–	–	–	–	–	–	2.0		6
	+	275	–	–	–	–	–	–	–	–	1.9		6
	+	275.5	0.806	2.829	24.05	0.352	24.2	20.06	–	–	1.94		3
	–	274	0.806	2.829	24.05	0.352	24.2	20.06	–	–	2.08		3
C$_3$H$_7$OH	+	271.5	0.821	3.073	25.3	0.346	24	25.66	25	3.7	3.0		2
i-C$_4$H$_9$OH	+	269.7	0.823	3.292	24.4	0.342	21.9	31.21	20	4.7	3.6		2
Isoamyl Alcohol	+	267.9	0.835	–	25.2	–	–	–	–	–	5.5		2
	+	271.1	0.835	–	25.2	–	–	–	–	–	4.1		1

Other Compounds

Compound	±												Ref.
H$_2$O	–	265	0.999	1.926	76.7	0.386	91.5	6.918	32	4.43	4.1		4
	–	283.1	0.9997	1.9258	74.22	0.386	84.11	6.8938	34	3.77	3.44		5
	–	297.8	0.99712	1.9275	72.02	0.386	78.65	6.8948	35	3.35	2.96		5
	–	323.6	0.9878	1.9335	67.82	0.386	69.78	6.9266	37	2.78	2.52		5
CHCl$_3$	+	258.9	1.58	3.105	31.8	0.210	5.74	18.34	25	5.8	3.0		2
	+	250	1.58	3.105	31.8	0.210	5.74	18.34	–	–	3.45		3
	–	249	1.58	3.105	31.8	0.210	5.74	18.34	–	–	3.6		3
C$_6$H$_6$	+	240	0.93	3.217	~36.7	0.000	2.274	9.922	15	17.0	5.4		2
	+	247	0.93	3.217	~36.7	0.000	2.274	9.922	–	–	4.95		3
	–	247	0.93	3.217	~36.7	0.000	2.274	9.922	–	–	4.98		3
CCl$_4$	+	237	~1.7	3.297	~33.85	0.000	2.226	10.40	15	15.9	5.7		3
	–	236	~1.7	3.297	~33.85	0.000	2.226	10.40	–	–	6.3		3
C$_6$H$_5$Cl	+	247	~1.2	3.337	–	0.352	5.77	22.82	–	–	10.2		3
	–	249	~1.2	3.337	–	0.352	5.77	22.82	–	–	8.9		3

[a] Calculated with $\gamma_{a'n} = \gamma_\infty [1 - (r_C/r_{a'n})^3]$ and $\varepsilon_{a'n+1}$ given by Eq. (198), $r_A = r_C$, $m_A = m_C$.
[b] *References:* 1. Laby (1908); 2. Przibram (1906); 3. Scharrer (1939); 4. see Table XVI; 5. Powell (1928); 6. Hertz (1956).

Considering these assumptions that it was necessary to introduce into the calculations, we find the agreement between the calculated and experimental values for the critical supersaturation to be excellent for H_2O and ethyl acetate, reasonable for the alcohols, and possibly acceptable for $CHCl_3$. The theory, however, apparently fails badly for two classes of compounds, the organic acids and nonpolar molecules.

For organic acids the model fails because it does not account for chemical interaction terms between molecules. The organic acids display such effects and have unusually large equilibrium concentrations of the n-mers, e.g., the concentration of the dimer of acetic acid often exceeds that for the monomer. In the absence of ions, our model for surface tension also failed for CH_3COOH, for the same reason (see Chapter IV).

With the nonpolar substances, the discrepancy may reflect inaccuracies in the experimental values. Loeb et al. (1938) found that the experimental value for C_6H_6 was very sensitive to impurities. For specially purified C_6H_6, the condensation temperature was not 240°K, but 215°K, which considerably raises the supersaturation needed for condensation, possibly to the point where homogeneous condensation occurs prior to heterogeneous condensation.

Figure 30 gives the theoretical curves for $S_{1,a\cdot n+1}$ for H_2O at 25 and 100°C on both H^+ and NH_4^+ ions. The ionic radii r_A were taken as 0 for H^+ and 2.18×10^{-8} cm for NH_4^+, the gas viscosity radius of NH_3. The curves indicate that the values for $S_{1,a\cdot n+1}$ decrease as the temperature is raised, except at very low n, where the reverse is true. They also indicate that $S_{1,a\cdot n+1}$ increases with ion size for small n, but this effect becomes less and less pronounced as n increases. For H_2O, $S_{1,a\cdot n+1}^{max} = 3.4$ at $n \sim 35$ at 25°C and $S_{1,a\cdot n+1}^{max} = 2.05$ at $n = 40$ at 100°C.

It is interesting to see how far this method can be pushed, i.e., to what size of smallness of the drop will such computations be valid. Of course as the drops fall below 10 molecules, they are no longer spherical, and the approximations may become quite bad. If, however, we do not demand too much from such a theory at $n < 10$, perhaps qualitative or semiquantitative results can be obtained.

The equilibrium constants $K_{1,a\cdot n}$ for the reaction

$$C + A^{\pm}C_n \rightleftarrows A^{\pm}C_{n+1}$$

are given by

$$-\ln(K_{1,a\cdot n}[C]_{vp}) = \exp\left\{\frac{4\pi}{\kappa T}\left[r_{a\cdot n+1}^2 \gamma_{a\cdot n+1} - r_{a\cdot n}^2 \gamma_{a\cdot n}\right] + \frac{\Delta H_{vap}}{\kappa T}\right\} \quad (203)$$

where a is the number of ions (1 in this case). The experimental values of $K_{1,a\cdot n}$ have been measured for $n \leq 7$ for H_2O with various ions in mass

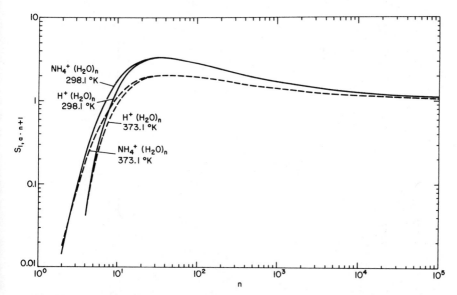

Fig. 30 Log–log plots of the supersaturation of C with $A_a \cdot C_{n+1}$, $S_{1,a \cdot n+1}$, versus n for solvated protons and ammonium ions at 298.1 and 373.1°K. The ionic radii were taken to be 0 and 2.18×10^{-8} cm, respectively.

spectrometer experiments, and they are listed in Table XIX. Also listed are the computed equilibrium constants from Eq. (203).

In the computations the ionic radii were taken to be 0 for H^+ and equal to the gas viscosity radius of NH_3, 2.18×10^{-8} cm for NH_4^+. For NO^+ and O_2^- ionic radii are not known but the internuclear distances are (Gilmore, 1965). We assumed that the ratio of the internuclear distance in O_2^- and O_2 was the same as the ratio of their corresponding gas viscosity radii. Similarly for NO^+ and NO. Thus the gas viscosity ionic radii were estimated to be 1.876×10^{-8} and 1.699×10^{-8} cm, respectively, for O_2^- and NO^+, and these values were used in the calculations.

The computed equilibrium constants reproduce the three important qualitative observations:

1. $K_{1,a \cdot n}$ decreases in any series as n increases.
2. $K_{1,a \cdot n}$ decreases as the temperature is raised.
3. $K_{1,a \cdot n}$ decreases as the size of the ion increases.

Experimentally condition 3 is not met if the sign of the ion changes, presumably because this also alters the surface tension, an effect which is unaccounted for by the theory.

It is apparent that the quantitative fit is not good for $H^+(H_2O)_n$ for $n < 4$, presumably because the droplets are not spherical. The agreement

Table XIX

Comparison of Calculated and Experimental Values of $K_{1,a\cdot n}$ (atm^{-1}) for the Reaction $H_2O + A^{\pm}(H_2O)_n \rightleftarrows A^{\pm}(H_2O)_{n+1}$

	$A^+ = H^+$ $10^8 r_A = 0$		$A^+ = NH_4^+$ $10^8 r_A = 2.18$ cm		$A^- = O_2^-$ $10^8 r_A = 1.876$ cm		$A^+ = NO^+$ $10^8 r_A = 1.699$ cm	
n	$K_{1,a\cdot n}\{calc\}^a$	$K_{1,a\cdot n}\{exp\}^b$	$K_{1,a\cdot n}\{calc\}^a$	$K_{1,a\cdot n}\{exp\}^c$	$K_{1,a\cdot n}\{calc\}^a$	$K_{1,a\cdot n}\{exp\}^d$	$K_{1,a\cdot n}\{calc\}^a$	$K_{1,a\cdot n}\{exp\}^e$
T = 298.1 K, ε_∞ = 78.54, γ_∞ = 71.97 dynes/cm, [Cl]$_{vp}$ = 0.01345 atm, r_C = 1.928 × 10^{-8} cm, q/e = 3.8588 × 10^{-9} cm, α = 6.8904 × 10^{-24} cm^3								
1	9.169 × 10^{11}	7.226 × 10^{17}	5.871 × 10^4	9.812 × 10^5	1.425 × 10^6	1.204 × 10^6	1.158 × 10^7	2.00 × 10^5
2	7.923 × 10^5	3.246 × 10^9	2.131 × 10^3	2.183 × 10^4	8.979 × 10^3	1.541 × 10^5	2.097 × 10^4	2.70 × 10^4
3	6.989 × 10^3	7.320 × 10^6	3.761 × 10^2	9.513 × 10^2				
4	7.235 × 10^2	1.239 × 10^4	1.336 × 10^2	1.645 × 10^2				
5	1.998 × 10^2	8.113 × 10^2						
6	8.912 × 10	1.285 × 10^2						
7	5.182 × 10	4.474 × 10						
T = 373.1 K, ε_∞ = 55.33, γ_C = 58.9 dynes/cm, [Cl]$_{vp}$ = 1.000 atm, r_C = 1.953 × 10^{-8} cm, q/e = 3.8588 × 10^{-9} cm, α = 7.06074 × 10^{-24} cm^3								
1	4.582 × 10^8	1.589 × 10^{13}	8.791 × 10^2	6.684 × 10^3				
2	5.856 × 10^3	4.336 × 10^6	5.251 × 10	2.312 × 10^2				
3	1.249 × 10^2	1.928 × 10^4	1.207 × 10	1.514 × 10				
4	1.948 × 10	6.883 × 10	5.010	6.117				
5	6.738	9.841						
6	3.443	2.424						
7	2.186	1.357						

[a] $K_{1,a\cdot n}\{calc\}$ from Eq. (203) with $\gamma_{a\cdot n} = \gamma_\infty [1 - (r_C/r_{a\cdot n})^3]$, ΔH_{vap} given by Eq. (196) and $\varepsilon_{a\cdot n}$ by Eq. (198).
[b] $K_{1,a\cdot n}\{exp\}$ from Grimsrud and Kebarle (1973).
[c] $K_{1,a\cdot n}\{exp\}$ from Payzant et al. (1973).
[d] $K_{1,a\cdot n}\{exp\}$ from Payzant and Kebarle (1972).
[e] $K_{1,a\cdot n}\{exp\}$ from Howard et al. (1971).

Table XX

Comparison of Calculated and Experimental Values of $K_{1,a\cdot n}(\text{atm}^{-1})$ for the Reaction $C + H^+ C_n \rightleftarrows H^+ C_{n+1}$ [a]

n	$K_{1,a\cdot n}\{\text{calc}\}$ [b]	$K_{1,a\cdot n}\{\text{exp}\}$	

$C = CH_3OH$, $T = 298.1°K$, $\varepsilon_\infty = 32.63$, $\gamma_\infty = 22.19$ dynes/cm, $[C]_{v.p.} = 0.1507$ atm, $r_C = 2.522 \times 10^{-8}$ cm, $q/e = 3.5526 \times 10^{-9}$ cm, $\alpha = 14.6528 \times 10^{-24}$ cm³ [c]

n	$K_{1,a\cdot n}\{\text{calc}\}$	$K_{1,a\cdot n}\{\text{exp}\}$
1	4.943×10^9	4.014×10^{17}
2	6.510×10^4	2.845×10^9
3	1.355×10^3	3.078×10^5
4	2.057×10^2	4.222×10^3
5	6.947×10	2.332×10^2
6	3.479×10	3.422×10
7	2.170×10	9.900

$C = CH_3OH$, $T = 373.1°K$, $\varepsilon_\infty = 17.46$, $\gamma_\infty = 15.67$ dynes/cm, $[C]_{vp} = 3.1720$ atm, $r_C \sim 2.60 \times 10^{-8}$ cm, $q/e = 3.5526 \times 10^{-9}$ cm, $\alpha \sim 13.60 \times 10^{-24}$ cm³ [c]

n	$K_{1,a\cdot n}\{\text{calc}\}$	$K_{1,a\cdot n}\{\text{exp}\}$
1	4.582×10^8	5.307×10^{12}
2	5.856×10^3	2.063×10^6
3	1.249×10^2	1.304×10^3
4	1.948×10	4.322×10
5	6.738	3.352
6	3.443	6.030×10^{-1}
7	2.186	1.686×10^{-1}

$C = NH_3$, $T = 298.1°K$, $\varepsilon_\infty = 16.9$, $\gamma_\infty = 20.2$ dynes/cm, $[C]_{vp} = 9.7961$ atm, $r_C = 2.061 \times 10^{-8}$ cm, $q/e = 2.9946 \times 10^{-9}$ cm, $\alpha = 7.36328 \times 10^{-24}$ cm³ [d]

n	$K_{1,a\cdot n}\{\text{calc}\}$	$K_{1,a\cdot n}\{\text{exp}\}$	
1	1.623×10^{11}	3.172×10^{12}	2.32×10^{11}
2	7.256×10^4	6.374×10^7	7.80×10^6
3	4.256×10^2	2.727×10^4	2.51×10^4
4	3.342×10	4.056×10^2	3.11×10^2
5	7.543	–	1.96
6	2.879	–	1.0

[a] Radius of $H^+ = 0$.
[b] $K_{1,a\cdot n}\{\text{calc}\}$ from Eq. (203), $\gamma_{a+n} = \gamma_\infty [1 - (r_C/r_{a\cdot n})^3]$, ΔH_{vap} from Eq. (196), and $\varepsilon_{a\cdot n}$ from Eq. (198).
[c] $K_{1,a\cdot n}\{\text{exp}\}$ from Grimsrud and Kebarle (1973).
[d] $K_{1,a\cdot n}\{\text{exp}\}$; first entry from Payzant et al. (1973); second entry from Arshadi and Futrell (1974).

improves as n increases, however, and becomes quite good for $n = 6$ or 7. The experimental investigators estimated their uncertainties to be about 10% in the free energies. At 25°C, this corresponds to an uncertainty in $K_{1,a\cdot n}$ of a factor of 60 at 10^{18} atm^{-1} and a factor of 2.0 at 10 atm^{-1}. At

100°C, the respective factors are 27 at 10^{18} atm^{-1} and 1.7 at 10 atm^{-1}. For $NH_4^+ (H_2O)_n$, the fit is good even for $n = 3$ or 4.

The theory can be further tested with other liquids. A comparison between computed and observed equilibria constants is given in Table XX for the reactions

$$H^+(CH_3OH)_n + CH_3OH \leftrightarrows H^+(CH_3OH)_{n+1}$$

and

$$H^+(NH_3)_n + NH_3 \leftrightarrows H^+(NH_3)_{n+1}$$

Again the fit is poor for $n < 5$, but reasonably good for $n > 4$.

Finally Table XXI shows a comparison for the reactions

$$A^-(CH_3CN)_n + CH_3CN \leftrightarrows A^-(CH_3CN)_{n+1}$$

where A^- represents the halide ions, and their radii are taken as their crystal radii. Again the fits are not too bad, considering the uncertainties in the experimental parameters and the limitations of the theory at low n. At least all the trends are proper; the equilibrium constants drop as n, r_A, and T increase.

Application to Crystals

The ideas outlined previously can be applied to crystal growth from solution. Glasner and Tassa (1974) have found that Pb^{2+} ions reduce the supersaturation needed to precipitate KBr and KCl. The crystals are formed in the same habit as in the absence of Pb^{2+} at solubilities of 4.1 and 3.1%, respectively, in excess of their saturation solubilities at 311 and 313°K, respectively. Furthermore they found that as the Pb^{2+} concentration increased, less of the salt precipitated until, at sufficiently high $[Pb^{2+}]$, no precipitation occurred at all. Presumably, as the $[Pb^{2+}]$ increases more and more potassium halide is tied up by the Pb^{2+} in smaller and smaller crystallites so that precipitation is more difficult. In fact, ultimately, enough potassium halide should be so tied up that the solution actually becomes undersaturated.

From the fact that small amounts of Pb^{2+} can precipitate the potassium halide, the size of the crystallite stable in a saturated solution can be estimated. For example, Glasner and Tassa found that a solution of $0.98 \times 10^{-5} M$ in Pb^{2+} at 36.5°C could precipitate KBr, but still leave the solution apparently supersaturated by $0.315 M$. Clearly this excess must be tied up in very small crystallites which were not collected. If all the Pb^{2+} are in the crystallites, each one contains 3.23×10^4 KBr ion pairs. The actual value may be somewhat smaller since some of the excess KBr may still be in solution. Likewise a solution $0.48 \times 10^{-5} M$ in Pb^{2+} at 37.6°C could

Table XXI

Comparison of Calculated and Experimental Values of $K_{1,a\cdot n}$ (atm^{-1}) for the Reaction $CH_3CN + A^-(CH_3CN)_n \rightleftarrows A^-(CH_3CN)_{n+1}$

	$A^- = F^-$ $10^8 r_A = 1.36$ cm		$A^- = Cl^-$ $10^8 r_A = 1.81$ cm		$A^- = Br^-$ $10^8 r_A = 1.95$ cm		$A^- = I^-$ $10^8 r_A = 2.16$ cm	
n	$K_{1,a\cdot n}\{calc\}^a$	$K_{1,a\cdot n}\{exp\}^b$	$K_{1,a\cdot n}\{calc\}^a$	$K_{1,a\cdot n}\{exp\}^b$	$K_{1,a\cdot n}\{calc\}^a$	$K_{1,a\cdot n}\{exp\}^b$	$K_{1,a\cdot n}\{calc\}^a$	$K_{1,a\cdot n}\{exp\}^b$
	T = 298.1°K, ε_∞ = 37.78, γ_∞ = 28.65 dynes/cm, $[C]_{vp}$ = 0.1168 atm, r_C = 2.757 × 10^{-8} cm, q/e = 7.996 × 10^{-9} cm, α = 19.3759 × 10^{-24} cm^3							
1	1.999 × 10^7	1.674 × 10^6	2.245 × 10^6	6.522 × 10^4	1.011 × 10^6	1.560 × 10^4	2.898 × 10^5	1.421 × 10^3
2	4.962 × 10^3	4.638 × 10^4	2.529 × 10^3	2.392 × 10^3	1.944 × 10^3	3.884 × 10^2	1.260 × 10^3	9.740 × 10
3	2.480 × 10^2	2.196 × 10^3	1.823 × 10^2	1.532 × 10^2	1.611 × 10^2	4.470 × 10		
4	5.673 × 10	1.856 × 10^2						
	T = 373.1°K, ε_∞ = 22.44, γ_∞ = 19.0 dynes/cm, $[C]_{vp}$ = 1.8703 atm, r_C = 2.884 × 10^{-8} cm, q/e = 7.996 × 10^{-9} cm, α = 21.0404 × 10^{-24} cm^3							
1	1.859 × 10^5	2.101 × 10^4	3.788 × 10^4	1.038 × 10^3	2.094 × 10^4	2.844 × 10^2	8.162 × 10^3	4.027 × 10
2	2.021 × 10^2	8.746 × 10^2	1.236 × 10^2	6.554 × 10	1.0174 × 10^2	1.304 × 10	7.356 × 10	4.147
3	1.676 × 10	6.436 × 10	1.336 × 10	1.869 × 10	1.219 × 10	6.913		
4	4.840	3.073 × 10						

$^a K_{1,a\cdot n}\{calc\}$ from Eq. (203) with $\gamma_{a\cdot n} = \gamma_\infty [1 - (r_C/r_{a\cdot n})^3]$, ΔH_{vap} from Eq. (196) and $\varepsilon_{a\cdot n}$ from Eq. (198).
$^b K_{1,a\cdot n}\{exp\}$ from Yamdagni and Kebarle (1972).

precipitate KCl, but still leave the solution apparently supersaturated by 0.230 M. The minimum size for the small crystallites must contain 4.96×10^4 KCl ion pairs.

From the considerations already developed, we can infer that for $n \gg 1$, the supersaturation equation for a cube is

$$\ln S_{1,a\cdot n+1} = -\frac{e\mathbf{q}}{\kappa T(\Upsilon' l)^2 \varepsilon_\infty} + \frac{6l_c^2 \gamma}{\kappa T}[(n+1)^{2/3} - n^{2/3}] \qquad (204)$$

where the surface energy term has been modified to reflect that the area is 6 times the cubic face area, and l_c^3 is the volume of one ion pair. The Coulombic term utilizes the fact that for crystals of any size $\varepsilon_{a\cdot n+1} = \varepsilon_\infty$. Since, however, the crystal is cubic, rather than spherical, the distance from the central Pb^{2+} ion to the edge $\Upsilon' l$ is variable, where l is the length of the cubic crystal. If deposition is occurring along a principal axis, then $\Upsilon' = \frac{1}{2}$, its minimum possible value. At the other extreme, i.e., along a diagonal, $\Upsilon' = \sqrt{3}/2$, its maximum value.

The surface term can be expanded by the binomial theorem, since $n \gg 1$, to give

$$\ln S_{1,a\cdot n+1} = -\frac{e\mathbf{q}}{\kappa T(\Upsilon' l)^2 \varepsilon_\infty} + \frac{4l_c^3 \gamma}{\kappa T l} \qquad (205)$$

Since the ion pairs separate in solution, the supersaturation is given by

$$S_1 = [K^+][X^-]/K_{sp} \qquad (206)$$

where K_{sp} is the solubility product for the particular crystal habit formed. Since, in the cases under consideration, Pb^{2+} does not alter the crystal habit, K_{sp} is just the usual solubility product. From the crystal spacings, both l_c and $\mathbf{q}$ are readily evaluated and the needed parameters are in Table XXII.

The critical size particle corresponds to the maximum in $S_{1,a\cdot n+1}$. Thus taking the derivative of Eq. (205) with respect to l, and setting it equal to zero gives

$$l_{max} = e\mathbf{q}/2l_c^3 \gamma \Upsilon'^2 \varepsilon_\infty \qquad (207)$$

Substitution of this expression into Eq. (205) permits us to compute

$$\gamma = (\kappa T e\mathbf{q} \ln S_{1,a\cdot n+1}^{max})^{1/2}/2\Upsilon' l_c^3 \varepsilon_\infty^{1/2} \qquad (208)$$

The values of γ_∞ are computed and included in Table XXII. The evaluation depends on the value of Υ' chosen, so that only limits to γ_∞ can be obtained. With the computed values of γ, l_{max} can be computed from Eq. (207), and these are also listed in Table XXII where they can be compared to the

Table XXII

Parameters for KBr and KCl Crystals

Compound	Υ'	KBr	KCl
T (°K)		311	313
ε_∞		2.33[a]	2.13[a]
q (esu)		1.580×10^{-17}	1.508×10^{-17}
$\ln S^{max}_{1,a \cdot n+1}$		0.0805	0.061
l_c^3 (cm^3)		71.9×10^{-24}	62.5×10^{-24}
γ (dyn/cm)	1/2	46.6	47.9
	$\sqrt{3}/2$	26.9	27.7
l_{max} (cm)	1/2	1.94×10^{-6}	2.27×10^{-6}
	$\sqrt{3}/2$	1.12×10^{-6}	1.32×10^{-6}
	Observed	$\lesssim 1.97 \times 10^{-6}$	$\lesssim 3.14 \times 10^{-6}$

[a] From Mott and Gurney (1953, p. 12).

observed values. The fit between the computed and observed values is quite satisfactory.

As more and more Pb^{2+} ions are put into solution, the small crystallites get smaller and smaller and the solution can become undersaturated. Since the crystallites are smaller, crystal growth by their coagulation is slowed and the bulk precipitate diminishes in quantity. Actually the coagulation coefficients are independent of the crystallite size, if the size is not too small, but more collisions are needed to produce the large size crystal that settles. In addition, as the crystallites get very small the rate coefficients decrease since each crystallite is charged and they repel each other. There is some minimum size when the charge repulsion energy e^2/l cannot be offset by the loss in surface free energy $2\gamma l^2$ on coalescence. Thus, below some minimum size, coagulation cannot occur and no precipitate will even form. This minimum size can be computed by equating the two energy terms giving

$$l^2 = (e^2/2\gamma)^{1/3} \tag{209}$$

For KBr and KCl the respective values for l computed from Eq. (209) are 11.7×10^{-8} cm ($n = 22.5$) and 11.8×10^{-8} cm ($n = 26.3$).

CHAPTER VII

ACCOMMODATION COEFFICIENTS

So far we have considered the problem of reaction assuming that every collision leads to reaction. In actual practice, even in the gas kinetic limit, the rate coefficients k_n or $k_{n,m}$ may be less than the collision coefficient, i.e.,

$$k_n = \alpha_n Z_m \tag{210}$$

$$k_{n,m} = \alpha_{n,m} Z_{n,m} \tag{211}$$

where α_n and $\alpha_{n,m}$ are the accommodation coefficients, respectively, for the reactions

$$C_n \rightleftarrows \text{physical source}$$

$$C_n + C_m \rightleftarrows C_{n+m}$$

The accommodation coefficients are ≤ 1.0. The forward and reverse reactions must have the same value for the accommodation coefficient in order to conform to the principle of microscopic reversibility.

TRANSMISSION FACTOR $\Gamma_{n,m}$

The arguments applicable to α_n are identical to those for $\alpha_{n,m}$, so we will consider only the problem for $\alpha_{n,m}$. This function is generally considered in the well-known Arrhenius form

$$\alpha_{n,m} = \Gamma_{n,m} p_{n,m} \exp\{-E_{n,m}/\kappa T\} \tag{212}$$

where $\Gamma_{n,m}$ is the transmission factor, $p_{n,m}$ the steric, or probability, factor, and $E_{n,m}$ the activation energy for the reaction.

Transmission Factor $\Gamma_{n,m}$

The transmission factor represents the fraction of successful reactions that do not return to reactants prior to equilibration as products. It is not to be confused with the reverse rate coefficient which gives the reverse reaction from equilibrated products. As an example of the distinction between the two processes, consider an egg thrown into a fan. First, the egg can react with the fan to equilibrium (i.e., scrambled egg), and then the reverse process can occur (i.e., the egg becomes unscrambled, replaced in the shell, and ejected). This is the true reverse reaction, which happens to be a very slow process in this example. Second the egg may pass through the fan without hitting the fan blades. There has actually been a collision in the gas kinetic sense, since the volume of the egg penetrated the volume of the fan, yet no reaction occurred, because the reactants were returned before equilibrium could occur. This corresponds to a transmission, and thus the transmission factor $\Gamma_{\text{egg,fan}}$ is < 1.0 (but not very much less, since almost all the time equilibrium will occur).

There are two reasons why $\Gamma_{n,m}$ may be < 1.0. One is for physical reasons and the other is for thermal reasons. Thus we consider $\Gamma_{n,m}$ as composed of two parts

$$\Gamma_{n,m} = \Gamma_{n,m}^p \Gamma_{n,m}^T \tag{213}$$

where $\Gamma_{n,m}^p$ is that due to physical limitations, and $\Gamma_{n,m}^T$ that due to thermal limitations.

Physical Transmission Factor $\Gamma_{n,m}^p$

Since $\Gamma_{n,m}^p$ is the same for the forward and reverse reaction, let us examine the reverse reaction

$$C_{n+m} \to C_n + C_m$$

Let us say that C_m is a small particle being ejected from the large particle, e.g., say $m = 1$. Thus as C crosses the surface of C_{n+1}, the reaction can be said to have occurred. If, however, there is an adsorbed layer or shell (often due to electrical forces) of gas phase (or solution phase) molecules enclosing C_n, C may not be able to escape, i.e., it may bounce off the adsorbed layer and return the pair $C_n - C$ into C_{n+1}. The net effect is that reaction has not occurred (in chemical kinetics this is referred to as the "cage effect"). Thus $\Gamma^p_{n,m}$ is < 1.0. This effect is very important in aerosol chemistry when the system is "dirty." The "dirt" attaches to the surface of the particles and the accommodation coefficients drop markedly, sometimes by a factor of 10 or more.

It should be realized that the higher the temperature, or more exactly, the greater the velocity of the ejected particle, the greater the likelihood to escape the "cage." Thus the accommodation coefficient can exceed the fraction of "free" surface of the particle.

This example was for a small particle ejected by a large particle. Another manifestation of the same problem occurs for two large particles. In this case consider the forward reaction

$$C_n + C_m \to C_{n+m}$$

If the two particles C_n and C_m that are approaching each other are moving in laminar flow fields, they each have a surface layer of the medium through which they are moving attached to them. As they approach each other they will deflect because of their fluid dynamic flow patterns, and no reaction will result. This might be considered a case in which collision never occurred. In the kinetic theory sense, however, collision has occurred because each particle felt the presence of the other one and changed its direction of motion as a result, i.e., a nonreactive collision occurred. This is the "cage effect" for the coalescence reaction.

In general, in gases which are "clean"

$$\Gamma^p_{n,m} = 1.0$$

As the system becomes "dirty" or the density and viscosity of the containing medium increase, $\Gamma^p_{n,m}$ drops.

Thermal Transmission Factor $\Gamma^T_{n,m}$

The thermal part of the transmission factor is of importance when the reaction complex separates before thermal equilibration. Thus the more energy in the reacting partners, the less likely it is that the reaction will occur (assuming that the activation energy barrier is overcome). Consider

STERIC FACTOR $p_{n,m}$

the egg encountering the spinning fan. The faster the egg is thrown, the more likely it is that it will pass through the fan intact. This is a well-known phenomenon in chemical kinetics—that collisions of very high energy react less often than those with energies just sufficient to pass the activation energy barrier.

A common example of a very low thermal transmission factor is that of the reaction of two atoms. They collide and separate in one vibrational time (of the diatomic molecule) because there is no place else to deposit the energy, unless a collision with a third particle occurs within the vibration time. The transmission factor is very low and is proportional to the pressure (in a gas). Thus atom–atom recombination is always a third-order process (in a gas).

Consider the reaction

$$C_n + C_m \rightleftarrows C_{n+m}^\dagger$$

where the superscript † indicates that the energy of reaction still resides in C_{n+m}. If $C_{n+m}^\dagger$ is formed, and the reverse process can occur before thermal equilibrium occurs, then $\Gamma_{n,m}^T < 1.0$. This is not considered forward followed by reverse reaction, which would be the case if $C_{n+m}^\dagger$ were first stabilized. In particle dynamics the "hot" $C_{n+m}^\dagger$ particle might be equilibrated, it might undergo the reverse reaction first, or it might decay by another route before thermal equilibration

$$C_{n+m}^\dagger \rightarrow C_{n+l} + C_{m-l}$$

In the last case, the overall reaction we have witnessed is

$$C_n + C_m \rightarrow C_{n+l} + C_{m-l}$$

This type of reaction was mentioned in the very first chapter and then discarded as unlikely. In fact in particle dynamics, there are usually so many degrees of freedom within the product particles that, except for impacts at extremely high velocities, thermal equilibration always occurs. For all practical purposes we can almost always say

$$\Gamma_{n,m}^T = 1.0$$

Steric Factor $p_{n,m}$

If two particles collide, reaction does not necessarily occur. The particles may "bounce" without crossing the reaction surface. One might consider the reverse "cage effect" discussed previously as an example of this. For symmetry reasons, however, we will consider the reverse "cage effect" a part of $\Gamma^p_{n,m}$, and reserve steric factors for two other reasons of nonreaction.

1. The angle of impact may influence the likelihood of reaction. We call this term $p^\phi_{n,m}$.

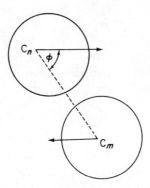

2. If the particles are not spherically symmetric, then the orientation on impact can influence the probability of reaction. We call this term $p^\sigma_{n,m}$. Consider the tetragonal particle colliding with a sphere. The impact may be along the short axis [case (a)], along the long axis [case (b)], or somewhere in between [case (c)]. The probability of reaction may be quite different for the different orientations.

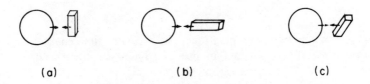

(a) (b) (c)

Angular Steric Factor $p^\phi_{n,m}$

In a random set of collisions, every direction in space is equally likely. Thus the probability of a collision occurring at any angle ϕ is proportional to the strip of area which the particles cross at that angle, i.e., $2\pi d \sin \phi$ for a given distance d. Thus the probability of a "head on" collision is zero, while that of a highly glancing collision increases with $\sin \phi$. Collisions that just graze are the most likely.

STERIC FACTOR $p_{n,m}$

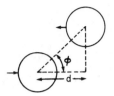

There are many reactions which have steric factors close to 1.0. For these reactions, the steric factor is not a function of impact angle (or more accurately, for large impact angles which account for most of the collisions, every collision is effective). There is no evidence that reduced steric factors for other reactions are in any way related to angle of impact. In fact they can always be related to orientation factors. There is no reason why particles should not behave the same way. For all situations we set

$$p_{n,m}^{\phi} = 1.0$$

This conclusion has some interesting implications when considering the reverse reaction which must behave in an identical way. When a particle C_{n+m} disintegrates into $C_n + C_m$, the probability is zero that they will separate along their line of centers. The reaction probability increases with $\sin \phi$, and the most probable reaction leads to separation as shown in the diagram. For angular momentum to be conserved, the particle C_{n+m} must have been rotating to begin with, or C_n and C_m must be rotating in a compensating way as they separate.

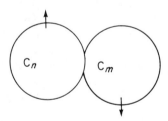

Orientation Steric Factor $p_{n,m}^{\sigma}$

If two particles collide, both of which are spherically symmetric, then there is no orientation factor, and $p_{n,m}^{\sigma}$ is 1.0. When asymmetry exists, a proper alignment is necessary for reaction to occur, and $p_{n,m}^{\sigma} < 1.0$. In general, for gas phase reactions where very specific alignments are needed for reaction, the values of $p^\sigma_{n,m}$ fall into two ranges:

$$p_{n,m} = 0.01-1.0 \quad \text{for noncyclic activated complexes}$$
$$p_{n,m} = 10^{-4}-10^{-2} \quad \text{for cyclic activated complexes}$$

Since coalescence presumably has a noncyclic activated complex, we can expect $p^\sigma_{n,m} > 0.01$. For specific particle shapes the values of $p^\sigma_{n,m}$ we might expect are listed in Table XXIII.

For liquids or amorphous solids, the particles are spherical, and at least for large particles, $(n > \sim 10)$, $p^\sigma_{n,m} = 1.0$. If the smaller particles are also spherical, then $p^\sigma_{n,m}$ will always be 1.0. For crystalline solids, the values of $p^\sigma_{n,m}$ may fall below unity because of their structured shape.

The steric factor $p^\sigma_{n,m}$ is determined by the orientation on impact. This may be different than the orientation during approach. If there is a strong attractive force between the particles (as there is at molecular distances), then the incoming particles will tend to become properly oriented during travel, further augmenting $p^\sigma_{n,m}$. On the other hand, the faster the particles approach (and rebound if there is no reaction) the less likely it is that the attractive force field will have time to reorient the particles and its influence will be diminished.

Activation Energy $E_{n,m}$

The activation energy is the energy the two particles must contain in order to pass the energy barrier to reaction as the reaction path is traversed. All coalescence reactions are exothermic, because the surface area is reduced, and the volume energy is unaffected (or reduced—depending on how the surface tension is computed for very small particles). Thus if a reaction path can be found which is energetically "downhill" all the way, $E_{n,m} = 0$.

A "downhill" reaction path exists if the surface area is continuously re-

Table XXIII

Expected Values of $p^\sigma_{n,m}$

C_n shape	C_m shape	$p^\sigma_{n,m}$
Spherical	Spherical	1.0
Spherical	Asymmetric	0.1–1.0
Asymmetric	Asymmetric	0.01–1.0

duced as the reaction proceeds. This will be the case for highly fluid materials (i.e., liquids). For more rigid particles this may not be the case, and for crystalline materials it will rarely be the case. If two very smooth and flat metal or crystalline surfaces collide, they will chemically bond. At the other extreme is the collision of two "billiard balls". For them to bond, rupture of at least one partner is necessary. This leads to an increase in surface area $\Delta A_{n,m}$ along the reaction path. Thus

$$E_{n,m} = \gamma \Delta A_{n,m} \tag{214}$$

Consider the case of two cubes colliding, each of radius r. Let the maximum fractional increment of area along the reaction path be Ψ. Then

$$E_{n,m} = \gamma \Psi 8\pi r^2 \tag{215}$$

The fractions of the collisions that can lead to reaction is

$$\exp\{-\gamma \Psi 8\pi r^2 / \kappa T\}$$

Room temperature results in the author's laboratory at Penn State have indicated that for crystalline NH_4NO_3 (Olszyna et al., 1974) and solid organic sulfur particles (Lauria et al., 1974a), the accommodation coefficient = 1.0 for $r < 3 \times 10^{-5}$ cm. On the other hand, the accommodation coefficient for particles of size $r > 10^{-2}$ cm must be near zero. Thus we can estimate Ψ. Taking a typical value of 40 dyn/cm for γ, we find that Ψ lies between 5×10^{-13} and 5×10^{-8}.

The value of Ψ of 5×10^{-8} corresponds to an absolute maximum value. If the particles are not of the same size, have larger surface-to-volume ratios (i.e., are not spherical), or are less rigid than crystals, then $\Psi < 5 \times 10^{-8}$ and, as in the case of liquids, will become zero.

Comparison with Experimental Values

Over the years, many determinations have been made for the accommodation coefficients for H_2O condensation (or evaporation). Until recently, these values usually fell into the range of 0.02 to 0.30, with most of the values near the lower bound.

Experimental values reported for accommodation coefficients obtained in evaporation are listed for H_2O and other liquids in Table XXIV. For a number of liquids the values are near 1.0, but some liquids give much lower values.

The difficulties with measurement of accommodation coefficients is that they all tend to give low values. The presence of dirt or impurities lowers

Table XXIV

Measured Accommodation Coefficients for Some Liquids from Evaporation Experiments

Liquid	Accommodation coefficient	Temp. (°C)	Reference
H_2O	0.04	14–19	Alty and Mackay (1935)
	>0.24	4.0	Hickman (1954)
	0.1, 0.24	—	Kappler and Hammick (1955)
	0.042	0	Delaney et al. (1964)
	0.027	43	Delaney et al. (1964)
	1.0[a]	11–23	Gollub et al. (1974)
	0.67[a,b]	18.8	Chodes et al. (1974)
C_2H_5OH	0.02	—	Baranaev (1939)
	0.024	12.4–15.5	Bucka (1950)
	0.036	0	Von Bogdandy et al. (1955)
CH_3OH	0.04	—	Baranaev (1939)
	0.017	32–35	Baer and McKelvey (1958)
	0.017	7	Delaney et al. (1965)
	0.022	−10	Delaney et al. (1965)
	0.030	−27	Delaney et al. (1965)
CCl_3H	0.18	—	Baranaev (1939)
Toluene	0.55	—	Baranaev (1939)
Benzene	0.9	—	Baranaev (1939)
Glycerol	0.052	18	Wyllie (1949)
	1.0	18–70	Trevoy (1953)
	0.05–0.15	13–25	Heideger and Boudart (1962)
Mercury	0.96	19.5	Knudsen (1915, 1916)
	1.0	−37 to +59	Volmer and Estermann (1921)
CCl_4	0.99	0	Von Bogdandy et al. (1955)
Dibutyl phthalate	1.0	15–35	Birks and Bradley (1949)
Boron	0.98	2130	Burns et al. (1967)
$n\text{-}C_{17}H_{36}$	0.95	22–40	Bradley and Shellard (1949)
$n\text{-}C_{18}H_{38}$	0.95	28–40	Bradley and Shellard (1949)
Potassium	0.95	66.7–119.3	Neumann (1954)
$SnCl_2$	0.96	350	Von Bogdandy et al. (1955)
Tridecylmethane	0.98	25–35	Bradley and Waghorn (1951)
Triheptylmethane	0.98	25–35	Bradley and Waghorn (1951)

[a] From condensation experiments.
[b] Recalculated from original data.

the measured value because part of the surface is blocked. Also if the drop is very small, then condensation liberates energy which heats the drop and increases the vaporation rate, thus giving an apparently reduced condensation rate. Similarly evaporation experiments cool the drop and thus reduce the evaporation rate.

In order to overcome these difficulties Chodes et al. (1974) made precise measurements of condensation rates on large drops under carefully controlled temperature conditions and supersaturations. From their theoretical treatment, they concluded that the accommodation coefficient was 0.033.

This value is, however, much too low as can be seen by examining their data. They found for supersaturations of 1.0072 at 18.8°C that H_2O drops of about 2.0×10^{-4} cm increased in radius at a rate of 0.22×10^{-4} cm/sec in air at 1 atm pressure. At this temperature, the vapor pressure of H_2O is 16.272 Torr.

From the expressions developed earlier we can compute the rate from the vapor pressure and the effective condensation coefficients since

$$k_{1,n+1}\{\text{eff}\} = k_{1,n}\left(1 - \frac{1}{S_1}\exp\left\{\frac{8\pi r_c^2 \gamma_\infty}{3\kappa T n^{1/3}}\right\}\right) \tag{216}$$

For drops of 2.0×10^{-4} cm, $n = 1.13 \times 10^{12}$ and $k_{1,n} = 3.0 \times 10^{-4}$ cm^3/sec if the accommodation coefficient is 1.0. The exponential term reduces to 1.0, and $k_{1,n+1}\{\text{eff}\}$ is easily computed. Then the computed growth rate in the volume of droplets of 2.0×10^{-4} cm radius is 50% greater than that obtained experimentally, so that the accommodation coefficient is 0.67. There is, however, sufficient experimental uncertainty in the temperature and supersaturation so that the accommodation coefficient could be 1.0.

Gollub et al. (1974) also measured the condensation coefficient of H_2O by measuring the rate of growth of water droplets at 11 to 23°C. They found an accommodation coefficient of 1.0 for particles of $(3-7) \times 10^{-4}$ cm radius at supersaturations of 1.02 to 1.05. For supersaturations <1.015, however, the accommodation coefficients were much lower, a result which the investigators could not explain. The results of Chodes et al. (1974) at a supersaturation of 1.0072 indicate, however, that even in this regime, the accommodation coefficient is near unity.

Accommodation coefficients for condensation and evaporation for a number of solids have been collected by Pound (1972b). The values for these coefficients all exceed 0.27, and most of them are near one. Experimental values have been obtained in three other systems in the author's laboratory at Penn State University for both condensation and coagulation coefficients. The accommodation coefficients have all been found to exceed 0.25 and probably 0.5, at least for particle sizes $< 2000 \times 10^{-8}$ cm radius. The three systems were quite different, both physically and chemically. They were:

1. formation of NH_4NO_3 crystals from the gas-phase reaction of NH_3 and HNO_3 (Olszyna et al., 1974);
2. formation of spherical solid particles of the composition $C_3H_4S_2O_3$ from the gas-phase photolysis of SO_2–C_2H_2 mixtures (Luria et al., 1974a).
3. formation of liquid drops of the composition C_5H_8SO from the gas-phase photolysis of SO_2–allene mixtures (Luria et al., 1974b).

Summary

In conclusion, we can expect that for clean liquids and small solid particles ($<10^{-5}$ cm), $\alpha_{n,m}$ should be close to unity. For larger crystalline solids, or if "dirt" (an adsorbed layer) is present, $\alpha_{n,m} < 1.0$, and in the case of the interaction of two solid particles each $>10^{-2}$ cm radius, $\alpha_{n,m}$ will approach zero.

REFERENCES

Allard, E. F., and Kassner, J. L., Jr. (1965). *J. Chem. Phys.* **42,** 1401. New Cloud-Chamber Method for the Determination of Homogeneous Nucleation Rates.

Allen, L. B., Jr. (1968). Ph.D. Dissertation, University of Missouri at Rolla. An Experimental Determination of the Homogeneous Nucleation Rate of H_2O Vapor in Ar and He.

Allen, L. B., and Kassner, J. L., Jr. (1969). *J. Colloid Interface Sci.* **30,** 81. The Nucleation of Water Vapor in the Absence of Particulate Matter and Ions.

Alty, T., and Mackay, C. A. (1935). *Proc. Roy. Soc.* **A149,** 104. The Accommodation Coefficient and the Evaporation Coefficient of Water.

Amelin, A. G. (1967). "Theory of Fog Condensation." S. Monson Wiener Bindery, Ltd., Jerusalem (English trans.).

Andrén, L. (1917). *Ann. Physik.* **52,** 1. Zählung und Messung der komplexen Moleküle einiger Dämpfe nach der neuen Kondensationstheorie.

Arshadi, M. R., and Futrell, J. H. (1974). *J. Phys. Chem.* **78,** 1482. Studies in High-Pressure Mass Spectrometry V Thermodynamics of Solvation Reactions. $NH_4^+-NH_3$.

Baer, E., and McKelvey, J. M. (1958). *A.I.Ch.E. J.* **4,** 218. Heat Transfer in Film Condensation.

Baranaev, M. (1939). *Zh. F. Kh.* **13,** 1635. Reaction between Surface Energy of Liquids and the Accommodation Coefficient.

Becker, R., and Döring, W. (1935). *Ann. Physik.* **24,** 719. Kinetische Behandlung der Keimbildung in übersättigten Dämpfen.

Biermann, A. H. (1971). Ph.D. Dissertation, University of Missouri at Rolla. Homogeneous Nucleation of Water Vapor in Inert Gas Atmospheres.

Birks, J., and Bradley, R. S. (1949). *Proc. Roy. Soc.* **A198,** 226. The Rate of Evaporation of Droplets II. The Influence of the Changes of Temperature and of the Surrounding Gas on the Rate of Evaporation of Drops of di-*n*-butyl phthalate.

Bolander, R. W., Kassner, J. L., Jr., and Zung, J. T. (1969). *J. Chem. Phys.* **50**, 4402. Semiempirical Determination of the Hydrogen Bond Energy for Water Clusters in the Vapor Phase I. General Theory and Application to the Dimer.

Bradley, R. S., and Shellard, A. D. (1949). *Proc. Roy. Soc.* **A198**, 239. The Rate of Evaporation of Droplets III. Vapor pressures and rates of evaporation of straight-chain paraffin hydrocarbons.

Bradley, R. S., and Waghorn, G. C. S. (1951). *Proc. Roy. Soc.* **A206**, 65. The Rate of Evaporation of Droplets V. Evaporation Characteristics of Some Branched-Chain Hydrocarbons and a Straight-Chain Fluorocarbon.

Bucka, H. (1950). *Z. Phys. Chem.* **195**, 260. Uber den Kondensationkoeffizienten von Äthylalkohol und ein Verfahren zur Bestimmung von Kondensation-koeffizienten.

Burns, R. P., Jason, A. J., and Inghram, M. G. (1967). *J. Chem. Phys.* **46**, 394. Evaporation Coefficient of Boron.

Chodes, N., Warner, J., and Gagin, A. (1974). *J. Atm. Sci.* **31**, 1351. A Determination of the Condensation Coefficient of Water from the Growth Rate of Small Cloud Droplets.

Collins, F. C. (1955). *J. Elektrochem.* **59**, 404. Time Lag in Spontaneous Nucleation due to Non-Steady-State Effects.

Countess, R. J., and Heicklen, J. (1973). *J. Phys. Chem.* **77**, 444. Kinetics of Particle Growth II. Kinetics of the Reaction of Ammonia with Hydrogen Chloride and the Growth of Particulate Ammonium Chloride.

Courtney, W. G. (1962a). *J. Chem. Phys.* **36**, 2009 Non-Steady-State Nucleation.

Courtney, W. G. (1962b). *J. Chem. Phys.* **36**, 2018. Kinetics of Condensation of Water Vapor.

Courtney, W. G. (1968). *J. Phys. Chem.* **72**, 421. Reexamination of Homogeneous Nucleation and Condensation of Water.

Delaney, L. J., Houston, R. W., and Eagleton, L. C. (1964). *Chem. Eng. Sci.* **19**, 105. Rate of Vaporization of Water and Ice.

Delaney, L. J., Psaltis, N. J., and Eagleton, L. C. (1965). *Chem. Eng. Sci.* **20**, 607. The Rate of Vaporization of Methanol.

de Pena, R. G., Olszyna, K. J., and Heicklen, J. (1973). *J. Phys. Chem.* **77**, 438. Kinetics of Particle Growth I. Ammonium Nitrate from the Ammonia–Ozone Reaction.

Flood, H. (1934). Ph.D. Dissertation. Institut fur Anorganische Chemie der Techn. Hochschule Norwegens, Trondheim, Norway. Tröpfchenbildung in Ubersättigten Dämpfen.

Frenkel, J. (1939). *J. Chem. Phys.* **7**, 538. A General Theory of Heterophase Fluctuations and Pretransition Phenomena.

Frenkel, J. (1955). "Kinetic Theory of Liquids," Chapter 7. Dover, New York.

Friedlander, S. K., and Wang, C. S. (1966). *J. Colloid Interface Sci.* **22**, 126. The Self-Preserving Particle Size Distribution for Coagulation by Brownian Motion.

Fuchs, N. A. (1964). "The Mechanics of Aerosols." Pergamon Press, Oxford (English transl.).

Gilmore, F. R. (1965). *J. Quant. Spectry. Rad. Transf.* **5**, 369. Potential Energy Curves for N_2, NO, O_2 and Corresponding Ions.

Glasner, A., and Tassa, A. (1974). *Israel J. Chem.* **12**, 799. The Thermal Effects of Nucleation and Crystallization of KBr and KCl Solutions Part II. The Heat of Nucleation and the Supersaturated Solution.

Gollub, J. P., Chabay, I., and Flygare, W. H. (1974). *J. Chem. Phys.* **61**, 2139. Laser Heterodyne Study of Water Droplet Growth.

Goodrich, F. C. (1964a). *Proc. Roy. Soc.* **A277**, 155. Nucleation Rates and the Kinetics of Particle Growth I. The Pure Birth Process.

Goodrich, F. C. (1964b). *Proc. Roy. Soc.*, **A277**, 167. Nucleation Rates and the Kinetics of Particle Growth II. The Birth and Death Process.

Grimsrud, E. P., and Kebarle, P. (1973). *J. Am. Chem. Soc.* **95**, 7939. Gas Phase Ion Equilibria

REFERENCES

Studies of the Solvation of the Hydrogen Ion by Methanol, Dimethyl Ether, and Water. Effect of Hydrogen Bonding.

Heicklen, J., and Luria, M. (1975). *Intern. J. Chem. Kinetics, Symp. No. 1*, p. 567. Kinetics of Homogeneous Particle Nucleation and Growth.

Heicklen, J., Hudson, J. L., and Armi, L. (1969). *Carbon* **7**, 365. Theory of Carbon Formation in Vapor Phase Pyrolysis-II. Variable Concentration of Active Species.

Heideger, W. J., and Boudart, M. (1962) *Chem. Eng. Sci.* **17**, 1. Interfacial Resistance to Evaporation.

Heist, R. H., and Reiss, H. (1973). *J. Chem. Phys.* **59**, 665. Investigation of the Homogeneous Nucleation of Water Vapor using a Diffusion Cloud Chamber.

Hertz, H. G. (1956). *Z. Elektrochem.* **60**, 1196. Investigations Using a Diffusion Cloud Chamber: Its Use in Measurement of Critical Supersaturation, and the Question of the Dependence of Critical Supersaturation on the Nature of the Carrier Gas.

Hickman, K. C. D. (1954). *Ind. Eng. Chem.* **46**, 1442. Maximum Evaporation Coefficient of Water.

Hidy, G. M., and Brock, J. R. (1970). "The Dynamics of Aerocolloidal Systems." Pergamon Press, Oxford.

Hirschfelder, J. O., Curtiss, C. F., and Bird, R. B. (1954). "Molecular Theory of Gases and Liquids." Wiley, New York.

Hirth, J. P., and Pound, G. M. (1963). *Prog. Mat. Sci.* **11**. Condensation and Evaporation, Nucleation and Growth Kinetics.

Homer, J. B., and Prothero, A. (1973). *Farad. Trans. I.* **69**, 673. Nucleation of Highly Supersaturated Vapors. Formation of Lead and Lead Oxide Smokes.

Howard, C. J., Rundle, H. W., and Kaufman, F. (1971). *J. Chem. Phys.* **55**, 4772. Water Cluster Formation Rates of NO^+ in He, Ar, N_2, and O_2 at 296°K.

Hudson, J. L., and Heicklen, J. (1967). *J. Phys. Chem.* **71**, 1518. The Relative Importance of Homogeneous and Heterogeneous Reactions.

Hudson, J. L., and Heicklen, J. (1968). *Carbon* **6**, 405. Theory of Carbon Formation in Vapor Phase Pyrolysis-I. Constant Concentration of Active Species.

Kantrowitz, A. (1951). *J. Chem. Phys.* **19**, 1097. Nucleation in Very Rapid Vapor Expansions.

Kappler, E., and Hammick, K. (1955). North Rhine West-Phalia Wirtschafts und Verkehrsmin. Forschungsber., No. 125, Eine neue Methode zur Bestimmg v. Kondensations Koeffizienten v. Wasser.

Kassner, J. L., Jr. (1975). *169th Am. Chem. Soc. Mtg., Philadelphia, Pennsylvania*. Vapor Phase Nucleation Rates Observed in an Expansion Chamber over a Wide Range of Temperatures.

Kassner, J. L., Jr., and Schmitt, R. J. (1966). *J. Chem. Phys.* **44**, 4166. Homogeneous Nucleation Measurements of Water Vapor in Helium.

Kassner, J. L., Jr., Plummer, P. L. M., Hale, B. N., and Biermann, A. H., (1971). Preprint from *Proc. Intern. Weather Modification Conf. Camberra, Australia, September 6–13*. The Role of Experiment in the Development of a Molecular Theory of Nucleation of Water Vapor— The Effects of Cluster Structure and Inclusion of Impurity Molecules.

Katz, J. L. (1970). *J. Chem. Phys.* **52**, 4733. Condensation of a Supersaturated Vapor I. The Homogeneous Nucleation of the n-Alkanes.

Katz, J. L., and Ostermier, B. J. (1967). *J. Chem. Phys.* **47**, 478. Diffusion Cloud-Chamber Investigation of Homogeneous Nucleation.

Katz, J. L., Scoppa, C. J., II, Kumar, N. G., and Mirabel, P. (1975). *J. Chem. Phys.* **62**, 448. Condensation of a Supersaturated Vapor II. The Homogeneous Nucleation of the n-Alkyl Benzenes.

Keyes, F. G. (1947). *J. Chem. Phys.* **15**, 602. The Thermodynamic Properties of Water Substance 0° to 150°C.

Kistenmacher, H., Lie, G. C., Popkie, H., and Clementi, E. (1974). IBM Research Report RJ 1334. Study of the Structure of Molecular Complexes VI. Dimers and Small Clusters of H_2O Molecules in the Hartree-Fock Approximation.
Knudsen, M. (1915). *Ann. Physik.* **47**, 697. Die Maximale Verdampflungsgeschwindigkeit des Quecksilbers.
Knudsen, M. (1916). *Ann. Physik.* **50**, 472. Die Verdichtung von Metalldämpfen an abgekühlten Körpern.
Knudsen, M., and Weber, S. (1911). *Ann. Physik.* **36**, 981. Luftwiderstand gegen die langsame Bewegung kleiner Kugeln.
Laby, T. H. (1908). *Phil. Trans. Roy. Soc.* **208**, 445. The Supersaturation and Nuclear Condensation of Certain Organic Vapors.
Leckenby, R. E., and Robbins, E. J. (1966). *Proc. Roy. Soc.* **A291**, 389. The Observation of Double Molecules in Gases.
Lindauer, G. C., and Castleman, A. W., Jr. (1971a). *Nucl. Sci. Eng.* **43**, 212. Initial Size Distributions of Aerosols.
Lindauer, G. C., and Castleman, A. W., Jr. (1971b). *Aerosol Sci.* **2**, 85. Behavior of Aerosols Undergoing Brownian Coagulation and Gravitational Settling in Closed Systems.
Loeb, L. B., Kip, A. F., and Einarsson, A. W. (1938). *J. Chem. Phys.* **6**, 264. On the Nature of Ionic Sign Preference in C. T. R. Wilson Cloud Chamber Condensation Experiments.
Lothe, J., and Pound, G. M. (1962). *J. Chem. Phys.* **36**, 2080. Reconsiderations of Nucleation Theory.
Luria, M., de Pena, R. G., Olszyna, K. J., and Heicklen, J. (1974a). *J. Phys. Chem.* **78**, 325. Kinetics of Particle Growth III. Particle Formation in the Photolysis of Sulfur Dioxide-Acetylene Mixtures.
Luria, M., Olszyna, K. J., de Pena, R. G., and Heicklen, J. (1947b). *Aerosol Sci.* **5**, 435. Kinetics of Particle Growth V. Particle Formation in the Photolysis of SO_2-Allene Mixtures.
Madonna, L. A., Sciulli, C. M., Canjar, L. N., and Pound, G. M. (1961). *Proc. Phys. Soc.* **78**, 1218. Low Temperature Cloud Chamber Studies on Water Vapor.
Mattauch, J. (1925). *Z. Physik* **32**, 439. Eine experimentelle Ermittlung des Widerstandsgesetzes kleiner Kugeln in Gasen.
Millikan, R. A. (1923). *Phys. Rev.* **22**, 1. The General Law of Fall of a Small Spherical Body through a Gas, and Its Bearing upon the Nature of Molecular Reflection from Surfaces.
Mott, N. F., and Gurney, R. W. (1953). "Electronic Processes in Ionic Crystals." Oxford Univ. Press, London and New York.
Neumann, K. (1954). *Z. Phys. Chem.* **2**, 215. Der Verdampfungskoeffizient des flüssigen Kaliums.
Olszyna, K. J., and Heicklen, J. (1972). *Adv. Chem. Ser.* **113**, 191. The Reaction of Ozone with Ammonia.
Olszyna, K. J., de Pena, R. G. Luria, M., and Heicklen, J. (1974). *Aerosol Sci.* **5**, 421. Kinetics of Particle Growth IV. NH_4NO_3 from the NH_3-O_3 Reaction Revisited.
Payzant, J. D., and Kebarle, P. (1972). *J. Chem. Phys.* **56**, 3482. Kinetics of Reactions Leading to $O_2{}^-(H_2O)_n$ in Moist Oxygen.
Payzant, J. D., Cunningham, A. J., and Kebarle, P. (1973). *Can. J. Chem.* **51**, 3242. Gas Phase Solvation of the Ammonium Ion by NH_3 and H_2O and Stabilities of Mixed Clusters $NH_4{}^+\text{-}(NH_3)_n(H_2O)_w$.
Pich, J., Friedlander, S. K., and Lai, F. S. (1970). *Aerosol Sci.* **1**, 115. The Self Preserving Particle Size Distribution for Coagulation by Brownian Motion-III. Smoluchowski Coagulation and Simultaneous Maxwellian Condensation.
Pound, G. M. (1972a). *J. Phys. Chem. Ref. Data* **1**, 119. Selected Values of Critical Supersaturation for Nucleation of Liquids from the Vapor.

Pound, G. M. (1972b). *J. Phys. Chem. Ref. Data* **1**, 135. Selected Values of Evaporation and Condensation Coefficients for Simple Substances.

Powell, C. F. (1928). *Proc. Roy. Soc.* **A119**, 553. Condensation Phenomena at Different Temperatures.

Probstein, R. F. (1951). *J. Chem. Phys.* **19**, 619. Time Lag in the Self-Nucleation of a Supersaturated Vapor.

Przibram, K. (1906). *Sitzwgbei d. kais Akad. d. Wissen in Wien, Mathnaturw. K2A* **115**, 33. Über die Kondensat. von Dampfen in Ionisierter Luft.

Reiss, H., and Katz, J. L. (1967). *J. Chem. Phys.* **46**, 2496. Resolution of the Translation–Rotation Paradox in the Theory of Irreversible Condensation.

Reiss, H., Katz, J. L., and Cohen, E. R. (1968). *J. Chem. Phys.* **48**, 5553. Translation–Rotation Paradox in the Theory of Nucleation.

Reiss, M. (1926). *Z. Phys.* **39**, 623. Die Beweglichkeit von Tröpfchen hober Dichte der Radiengrössen bis 1.10^{-5} cm und deren elektrische Ladungen.

Rowlinson, J. S. (1949). *Trans. Farad. Soc.* **45**, 974. The Second Virial Coefficients of Polar Gases.

Sander, A., and Damköhler, G. (1943). *Naturwissen.* **31**, 460. Ubersättigung bei der spontanen Keimbildung in Wasserdampf.

Scharrer, L. (1939). *Ann. Physik.* **35**, 619. Kondensation von übersättigten Dämpfen an Ionen.

Schmitt, K. H. (1959). *Z. Naturforsch* **14A**, 870. Untersuchungen an Schwebstoffteilchen in Temperaturfeld.

Smoluchowski, M. (1916). *Phys. Zeit.* **17**, 557 and 585. Drie Vortrage uber Diffusion, Brownsche Molekularbewegung und Koagulation von Kolloidteilchen.

Tolman, R. C. (1949). *J. Chem. Phys.* **17**, 333. The Effect of Droplet Size on Surface Tension.

Trevoy, D. J. (1953). *Ind. Eng. Chem.* **45**, 2366. Rate of Evaporation of Glycerol in High Vacuum.

Twomey, S. (1959). *J. Chem. Phys.* **31**, 1684. Nucleation of Ammonium Chloride Particles from Hydrogen Chloride and Ammonia in Air.

Volmer, M. (1939). *Kinetik der Phasenbildung.* Steinkopff, Dresden and Leipzig.

Volmer, M., and Estermann, I. (1921). *Z. Phys.* **7**, 1. Über den Verdampfungskoeffizienten von festem und flüssigem Quecksilber.

Volmer, M., and Flood, H. (1934). *Z. Phys. Chem.* **A170**, 273. Tröpfchenbildung in Dämpfen.

Volmer, M., and Weber, A. (1926). *Z. Phys. Chem.* **119**, 277. Keimbildung in übersättigten Gebilden.

Von Bogdandy, L., Kleist, H. G., and Knacke, O. (1955). *Z. Elektrochem.* **59**, 460. Velocity of Evaporation of Ethanol, CCl_4, and Sn(II) Chloride.

Wakeshima, H. (1954). *J. Chem. Phys.* **22**, 1614. Time Lag in the Self-Nucleation.

Wang, C. S., and Friedlander, S. K. (1967). *J. Colloid Interface Sci.* **24**, 170. The Self-Preserving Particle Size Distribution for Coagulation by Brownian Motion II. Small Particle Slip Correction and Simultaneous Shear Flow.

White, D. R., and Kassner, J. L., Jr. (1971). *Aerosol Sci.* **2**, 201. Experimental and Theoretical Study of the Sign Preference in the Nucleation of Water Vapor.

Wilson, C. T. R. (1897). *Phil. Trans. Roy. Soc.* **A189**, 265. Condensation of Water Vapour in the Presence of Dust-free Air and Other Gases.

Wilson, C. T. R. (1900). *Phil. Trans. Roy. Soc.* **193**, 289. On the Comparative Efficiency as Condensation Nuclei of Positively and Negatively Charged Ions.

Wyllie, G. (1949). *Proc. Roy. Soc.* **A197**, 383. Evaporation and Surface Structure of Liquids.

Yamdagni, R., and Kebarle, P. (1972). *J. Am. Chem. Soc.* **94**, 2940. Solvation of Negative Ions by Protic and Aprotic Solvents. Gas Phase Solvation of Halide Ions by Acetonitrile and Water Molecules.

INDEX

A

Acetic acid, 74–76, 100, 102, 104
   dimers, 55
Acetone, 100
Activation energy, 118–119
Alkyl benzenes, 77
Allene, 82–83, 121
Ammonia, see NH_3
Ammonium, see NH_4
Aniline, 100
Application to crystals, 108–111
   supersaturation, 110
Ar, 71, 72
Average velocity, 15

B

$BaHgI_4$ diffusion coefficient, 20
Becker and Döring theory, 63–65, 71, 76–79
Benzene, 99–100, 103–104, 120
   accommodation coefficient, 120
Boron, 120
   accommodation coefficient, 120
$Br^-(CH_3CN)_n$, 108–109
Butyric acid, 74

C

CCl_4, 99, 103, 120
   accommodation coefficient, 120
$CClH_3$, 48–49, 100
CCl_3H, 100, 103–104, 120
   accommodation coefficient, 120
CH_3CHO, 49, 59
CH_3CN, 49–50
CH_3NO_2, 74
CH_3OH, 74, 76–77, 100, 102, 107, 120
   accommodation coefficient, 120
CH_3COOH, 100, 102
C_2H_2, 79–82, 121
C_2H_5COOH, 102

C_2H_5I, 100
C_2H_5OH, 74, 76–77, 100, 120
   accommodation coefficient, 120
$C_3H_4S_2O_3$, 78–79, 121
$(C_3H_4S_2O_3)_3$, 61
C_3H_7COOH, 102
C_3H_7OH, 74, 103
C_4H_9Br, 100
C_4H_9OH, 74, 103
C_5H_8SO, 61, 121
C_6H_5Cl, 100, 103
$C_6H_5NO_2$, 100
C_6H_6, 99–100, 103–104, 120
$C_{17}H_{36}$, 120
   accommodation coefficient, 120
$C_{18}H_{38}$, 120
   accommodation coefficient, 120
CO_2, 51–53
Cage effect, 114
Chemical source term, 2–3
$Cl^- \cdot (CH_3CN)_n$, 108–109
Coagulation, 2–4, 34–39
   log–normal distribution, 37–39
Coalescence, 2–4
Condensation, 4, 26–34, 90–92
   constant [C], 26–32
   effective coefficients, 43
   rate of, 42
   on soluble condensation nuclei, 90–92
   variable [C], 32–34
Condensation nuclei, insoluble, 92–96
Condensation nuclei, soluble, 87–92
   condensation, 90–92
   nucleation, 88–89
   supersaturation, 91–94

D

Deposition, net rate of, 42
Detailed balancing, law of, 48
Dibutyl phthalate, 120
   accommodation coefficient, 120
Dielectric constant, 97–98
Diffusion coefficient, 17–20
   $BaHgI_4$, 20
   glass spheres, 20
   oil droplets, 19–20
   silicone oil, 20
Diffusional loss, 5–14
   non-steady-state solution, 13–14
   pure diffusional loss, 13

simplified solution, 7–8
steady-state solution, 8–13
   $K_v = 0$, 11
Diluent, effect of on homogeneous nucleation, 71–72
Dipole moment, 97
Disintegration, 2–4

E

Enthalpy of vaporization, 93–94, 97–98
Equilibrium constant, 40–44, 50–55, 106–107
   $Br^-(CH_3CN)_n$, 108–109
   $Cl^-(CH_3CN)_n$, 108–109
   $F^-(CH_3CN)_n$, 108–109
   $I^-(CH_3CN)_n$, 108–109
Equilibrium theory of homogeneous nucleation, 57–63, 67
   constant critical size q, 59–60
   induction time, 62–63
   variable critical size q, 60–62, 74
   Volmer–Weber theory, 61
Ethyl acetate, 74, 102
Ethyl propionate, 74, 102
Evaluation of parameters, 15–20
   average velocity, 15
   collision rate coefficient, 15
   diffusion coefficient, 17–20
   mean free path, 16, 18
   reduced mass, 16
   viscosity, 16
Expansion ratios, 100

F

$F^-(CH_3CN)_n$, 108–109
First-order loss processes, 6
Formic acid, 74, 102
Fundamental processes, 1–4

G

Glass spheres, diffusion coefficient of, 20
Glycerol, 120
   accommodation coefficient, 120
Gold, 71, 72
Gravitational settling, 14–15
   settling rate coefficient, 15
   settling velocity, 15
   spatial distribution, 14–15

INDEX

H

H^+, 105
$H^+(CH_3OH)_n$, 107–108
HCOOH, 74, 102
HCl, 82, 85–86
$H^+(H_2O)_n$, 105–106
$H^+(NH_3)_n$, 107–108
HNO_3, 82–85, 121
H_2, 76, 79, 80
H_2O, 48–55, 59–62, 64–65, 67–74, 76–77, 92, 98–101, 103–104, 119–121
    accommodation coefficient, 120
He, 71–73, 76–78
Heptane, 77, 79
Hexane, 77–78
Hg, 120

I

$I^-(CH_3CN)_n$, 108–109
Induced dipole, 97
Induction time, 62–63, 66–68
Ions as condensation nuclei, 97–113
    application to crystals, 108–111
    comparison with experiments, 98–108
    expansion ratios, 100
    supersaturation, 104–105
Isoamyl alcohol, 103
Isobutyric acid, 74
Isovaleric acid, 74, 102

K

K, 120
$k_n\{wall\}$, 6–7
KBr, 108–111
KCl, 108–111
Kinetic theory of homogeneous nucleation, 68–70
    [C] constant, 69
    constant total mass, 70
Kr, 72

L

Lead, 61
Lead ion, 108–111
Log–normal distribution, 37–39
Lothe–Pound theory, 59, 79–80
Low n, results at, 48–56
    acetic acid, 55
    equilibrium constants, 50–55
    H_2O, 50–55
    supersaturation, 52–56
    surface energy (tension), 50, 55
    Tolman's formula, 50, 55

M

Mean free path, 16, 18
Mercury, 120
    accommodation coefficient, 120
Methyl butyrate, 74, 102
Methyl isobutyrate, 74, 102

N

NH_3, 49, 52, 82–86, 105, 107
NH_3–HCl, 85–86
NH_4^+, 105
NH_4Cl, 85–86
$(NH_4Cl)_q$, 61
$NH_4^+(H_2O)_n$, 105–106, 108
NH_4NO_3, 83–85, 119, 121
$(NH_4NO_3)_q$, 61
NO, 105
NO^+, 105
$NO^+(H_2O)_n$, 106
Ne, 72
Nonane, 77
Nucleation, 4, 57–72, 88–89, 96
    effect of diluent on, 71–72
    equilibrium theory of, 57–63
    kinetic theory of, 68–70
    on insoluble condensation nuclei, 96
    on soluble condensation nuclei, 88–89
    steady-state theory of, 63–68
Nucleation from chemical reaction, 79–86
    NH_3–HCl, 85–86
    NH_3–HNO_3, 83–85
    SO_2^*–allene, 82–83
    SO_2^*–C_2H_2, 79–82
    $C_3H_4S_2O_3$, 78–79

O

O_2, 105
O_2^-, 105
$O_2^-(H_2O)_n$, 106
Octane, 77, 80
Oil droplets, diffusion coefficient of, 19–20

P

Pb, 61
Pb^{2+}, 108–111
Physical removal and source terms, 2–4
Poisson process, 30–31
Polarizability, 97, 101
Potassium, 120
   accommodation coefficient, 120
Propionic acid, 74
Propyl acetate, 74, 102

R

Radius, 41
   apparent molecular, 41
   equivalent spherical, 41
Rate coefficient, 21–25, 42–45, 112
   collision, 15, 21–22, 24–25
   collision effective, 42, 44
   deposition, rate of, 44
   influence of diffusion, 22–25
   wall rate removal speed, 45
Reduced mass, 16
Reiss–Katz–Cohen theory, 59, 79–80

S

SO_2, 79–83, 121
Settling rate coefficient, 15
Settling velocity, 15
Silicone oil, diffusion coefficient of, 20
$SnCl_2$, 120
   accommodation coefficient, 120
Spontaneous fracture, 2–4, 43–45
   equilibrium constant, 43–44
   rate coefficient, 44–45
Steady-state theory of homogeneous nucleation, 63–68
   Becker and Döring theory, 63–65
   modified steady-state theory, 65–66, 74
   induction time, 66–68
Steric factor, 115–118
   angular, 116–117
   orientation, 117–118
Stokes' Law, 17
Supersaturation, 45–48, 52–56, 70–71, 73, 76–80, 91–94, 98–99, 102, 104–105, 110
   critical, for homogeneous nucleation, 72–78
   H_2O, 46, 48
   steady-state, 45–48, 52–56, 93–96, 98
Surface energy (tension), 41, 50–55

T

Thompson energy, 97, 98
Tolman's formula, 50, 55
Toluene, 100, 120
   accommodation coefficient, 120
Transmission factor, 113–115
   physical, 113–114
   thermal, 114–115
Tridecylmethane, 120
   accommodation coefficient, 120
Triheptylmethane, 120
   accommodation coefficient, 120

V

Vaporization, 4, 40–43
   enthalpy of, 93–94, 97–98
   equilibrium constant, 40–42
   rate coefficient, 42–43
Vaporization rate, 42
Viscosity, 16
Volmer–Weber Theory, 61

W

Wall rate removal speed, 43
Wilson cloud chamber, 72, 98–100

X

Xe, 72